Targeting Economic Incentives for Environmental Protection

MIT Press Series on the Regulation of Economic Activity

General Editor
Richard Schmalensee, MIT Sloan School of Management

1 *Freight Transport Regulation*, Ann F. Friedlaender and Richard H. Spady, 1981

2 *The SEC and the Public Interest*, Susan M. Phillips and J. Richard Zecher, 1981

3 *The Economics and Politics of Oil Price Regulation*, Joseph P. Kalt, 1981

4 *Studies in Public Regulation*, Gary Fromm, editor, 1981

5 *Incentives for Environmental Protection*, Thomas C. Schelling, editor, 1983

6 *United States Oil Pipeline Markets: Structure, Pricing, and Public Policy*, John A. Hansen, 1983

7 *Folded, Spindled, and Mutilated: Economic Analysis and* U.S. *v.* IBM, Franklin M. Fisher, John J. McGowan, and Joen E. Greenwood, 1983

8 *Targeting Economic Incentives for Environmental Protection*, Albert L. Nichols, 1984

Targeting Economic Incentives for Environmental Protection

Albert L. Nichols

The MIT Press
Cambridge, Massachusetts
London, England

This book was set in Times New Roman by Asco Trade Typesetting Ltd., Hong Kong and printed and bound by Halliday Lithograph in the United States of America.

Library of Congress Cataloging in Publication Data

Nichols, Albert L.
  Targeting economic incentives for environmental protection.

  (MIT Press series on the regulation of economic activity; 8)
  Bibliography: p.
  Includes index.
  1. Environmental policy—Cost effectiveness.   I. Title.   II. Series.
HC79.E5N47   1984   338.4′33637   83-24883
ISBN 0-262-14036-5

To My Mother

Contents

List of Tables

List of Figures

Series Foreword

Government regulation of economic activity in the United States has grown dramatically in this century, radically transforming government-business relations. Economic regulation of prices and conditions of service was first applied to transportation and public utilities and was later extended to energy, health care, and other sectors. In the early 1970s explosive growth occurred in social regulation, focusing on workplace safety, environmental preservation, consumer protection, and related goals. Though regulatory reform has occupied a prominent place on the agendas of recent administrations, and some important reforms have occurred, the aims, methods, and results of many regulatory programs remain intensely controversial.

The purpose of the MIT Press series, Regulation of Economic Activity, is to inform the ongoing debate on regulatory policy by making significant and relevant research available to both scholars and decision makers. Books in this series present new insights into individual agencies, programs, and regulated sectors, as well as the important economic, political, and administrative aspects of the regulatory process that cut across these boundaries.

Most economic models of environmental problems deal with a simple world in which "emissions" and "damages" are simply two names for the same easily measured, one-dimensional quantity. In such a world the economic case for an incentives approach to environmental protection is strong, and the design of efficient incentive schemes is straightforward. But in the real world, emissions and damages are vector valued, vary across time and space, are not costless to monitor, and are not always linked in simple ways. In this world, the design of efficient incentive schemes is generally not straightforward, may be important, but has been traditionally neglected by economists. In this volume Albert Nichols tackles this problem. Weaving theoretical arguments and the careful analysis of a significant case study, he stresses the importance of careful choice of the targets at which an incentive scheme (or other regulatory device) is aimed. His findings are of immediate applicability to the design of policies for air quality control; his arguments and findings have broader implications for the design of regulatory policies and rules.

Economists have long argued that environmental regulation should rely more on economic incentives and less on direct control through uniform standards. This study began as part of a larger project designed to explore some of the practical and political obstacles to the implementation of incentive-based approaches to environmental protection. My role was to prepare a case study of airborne benzene, focusing on the advantages and disadvantages of using charges or permits rather than the more usual uniform emission standards. (The results of the project, including the benzene case study, appear in Schelling 1983.) That project helped convince me that incentives were practical and efficient mechanisms. It also convinced me, however, that by focusing almost exclusively on variations in the costs of controlling emissions, students and practitioners of regulation had missed the significance of large variations in the marginal benefits of control.

Emission charges, marketable emission permits, and other traditional incentive schemes are designed to allocate control efforts where they will yield the largest reductions in emissions. The goal of environmental regulation, however, is to reduce damages, not emissions per se. In the case study of benzene, it quickly became apparent that it mattered not only by how much emissions were reduced but also where those reductions occurred. For one class of sources (maleic anhydride plants), reducing emissions at one plant by 1 kg of benzene provided the same benefits as reducing them by 50 kg at another. For other categories the variations were even larger. These wide variations in the marginal benefits of control suggested the need to develop strategies that were sensitive to those variations, as well as to differences in marginal costs.

The benzene case study also raised the issue of what to do when damages are a function of several different variables. The ideal is to state the regulation in terms of a single performance measure, allowing each polluter to choose the least-cost combination of actions. Often, however, it is difficult to monitor and enforce performance-based regulations, so separate standards (or incentives) must be used. The Environmental Protection Agency, for example, considered a standard limiting the benzene content of gasoline and another requiring that service stations install equipment to control gasoline vapors that contain benzene (and other substances). The general issue that interested me was how much efficiency is sacrificed when regulations are aimed at the specifications of individual components, rather than at overall performance measures.

These issues are examples of the more general problem of what I call "choosing the target," selecting the specific point(s) at which a standard or incentive will be imposed. This book shows how the choice of target interacts with the choice of instrument to determine the efficiency of environmental regulations. Although originally stimulated by my work on benzene, virtually all of the analysis applies to a wide range of environmental issues, and much of it has implications for the design of regulatory policies in other areas as well.

Acknowledgments

This book is based on my doctoral dissertation, completed in 1981. I am grateful to several organizations for providing financial support at various stages. The Environmental Protection Agency supported the original case study of benzene. The National Science Foundation and the Harvard Faculty Project on Regulation provided support while I was working on the dissertation. The Regulation Project, with funding from the General Electric Foundation, also supported revisions for this book and preparation of the final manuscript.

I am also grateful to Sally Makacynas, Barbara Nielson, and Arlene Pippin for their efforts in typing various drafts of the manuscript. I owe a special debt to Holly Grano, who typed early drafts of some chapters, intermediate drafts of many, and most of the final manuscript. Her speed, accuracy, and unflappable good humor were especially welcome in the final push to finish the manuscript.

The three members of my thesis committee—Howard Raiffa, Thomas Schelling, and Richard Zeckhauser—have contributed greatly to my thinking on the topics covered in this book, both through their advice and their writings. Professor Schelling provided particularly helpful and detailed comments on the benzene case study. My greatest debt is to Professor Zeckhauser, who has been a continuing source of intellectual stimulation and sound advice, first as my principal advisor and more recently as a colleague. Alan Carlin and Richard Johnson of the Environmental Protection Agency were both generous in taking the time to guide me to information on the benzene case study. Jeffrey Harris, David Harrison, William Hogan, and Richard Schmalensee all provided helpful suggestions for revising the manuscript.

My wife, Eve, deserves more than the usual thanks accorded to spouses. In addition to shouldering more than her fair share of family responsibilites so that I could work on the thesis and later this book, she also edited earlier drafts of all of the chapters. She and our children, Matthew and Beth, also provided one of the stronger (though noneconomic) incentives for me to finish, so that I might spend more time with them instead of working on "Daddy's book."

Chapter 1

Introduction

In April 1980 the U.S. Environmental Protection Agency (EPA) proposed a regulation to limit the emissions of benzene from chemical plants that use it as feedstock to produce maleic anhydride (45 *Fed. Reg.* 26660 1980). The proposal was the first in a planned series of emission standards for benzene, a suspected cause of leukemia. It was also part of a larger effort to control airborne carcinogens.

The impact of the standard would have been modest, at least when judged in comparison with many other environmental regulations. EPA estimated that its annual cost would be just over $2 million and that it would eliminate a death from leukemia about once every two and a half years.[1] The proposal, however, embodied many of the characteristics and failings of current U.S. policy toward environmental pollutants generally, and toward suspected carcinogens in particular. The standard proposed was not based on a careful comparison of the costs and benefits of alternative control levels but rather reflected the "best available technology" (BAT), that is, the tightest standard that could be imposed within the limits of existing technology without causing the closing of many plants and the loss of a significant number of jobs. New plants would not be allowed to emit any benzene, and existing plants were to achieve approximately 97 percent control.

The proposed standard also was typical in that it relied on what Charles Schultze (1977) has called the "command and control" approach. It specified how much benzene each plant could emit, with a uniform rate for all existing plants. No account was taken of differences across plants in the marginal cost of controlling emissions, although those costs vary widely.

The uniform standard proposed also failed to take account of interplant differences in the benefits of controlling emissions, although some plants are located in densely populated urban areas, whereas others are at remote rural sites where few people are affected by their emissions, so few people would benefit from control. For these reasons the proposed regulation was inefficient, commanding a pattern of controls far more costly than necessary to yield the health benefits that the standard would achieve.

The Economic Critique of Environmental Regulation

Economists critical of current approaches to environmental protection usually emphasize two major faults.[2] First, regulators rarely make much effort to compare the costs and benefits of alternative standards; they almost never use benefit-cost or cost-effectiveness analyses. Indeed much environmental legislation appears to deny that the trade-off between protection and control costs should be considered at all, except in the most extreme cases such as potential plant closures. Implicit trade-offs are made but in ad hoc and often arbitrary ways that lead invariably to inconsistencies across regulations and agencies. The level of control defined by BAT for one type of plant, for example, may cost $100 million per fatality averted, whereas a BAT standard for another category may have a cost of only $100,000 per life saved. Whatever one's belief about the amount that should be spent to protect health, such inconsistencies are inefficient; they mean that adjustments in the stringencies of standards could both increase protection and reduce costs.

The second major criticism is that direct regulation through standards is inherently inefficient, even if the standards are based on the best scientific evidence available and the most careful economic analysis possible. The central problem is that standards attempt to impose rigid, uniform solutions on heterogeneous firms and individuals in a highly uncertain and rapidly changing world. The outcome is a highly inefficient allocation of control efforts across emission sources, too tight in some cases and too lenient in others.

Incentive-Based Alternatives

The alternative long favored by economists for controlling pollution and other externalities is to use financial incentives, such as emission charges or marketable emission permits. The basic prescription dates back at least half a century, to Pigou's (1932) *Economics of Welfare*. Pigou proposed that when an economic activity generates external costs, the government should impose a tax on that activity equal to those external costs, thus internalizing those costs and providing an incentive for their reduction. Since that time, particularly over the past decade or two, as pollution and other externalities have become recognized as major problems in industrialized economies, Pigou's prescription has been elaborated and modified, but the core idea remains unchanged: externalities are best dealt with by harnessing market forces, rather than by attempting to override the market through the imposition of direct regulation.[3]

The rationale for using incentive-based mechanisms is straightforward. The market misallocates resources in the face of negative externalities because individual economic actors do not bear the full costs of their actions. Firms emitting carcinogens, for example, do not bear the costs of the cancers caused among nearby residents, except in those rare instances where the link between emissions and individual cases of illness is so clear that the victim can bring a successful tort action. Thus the firm has little incentive to control its emissions. By imposing a charge equal to the divergence between private and social marginal costs, the government provides the appropriate incentive for the firm to reduce its hazardous activity. Alternatively, the government can create a new system of property rights that internalizes costs by forcing firms to purchase permits to emit.[4]

Under either type of incentive scheme the emphasis is on decentralized decision making about particular modes and levels of control, with the choices made by those most familiar with the circumstances at individual plants. In contrast, standards impose a centralized and often uniform decision by the regulatory agency, which has limited knowledge of the industry affected and little ability to tailor regulations to source-specific conditions. The primary virtue of incentive schemes is that they allocate control efforts in accordance with the marginal costs of control; expenditures are made where they will yield the largest reductions in emissions. When appropriately structured, they can achieve environmental goals at minimum cost. Strong theoretical arguments in favor of incentives have been buttressed by a limited number of empirical investigations that suggest large potential savings over current approaches.[5]

Resistance to Incentives
Despite the apparent advantages of incentives, until very recently Congress and most regulators have shown virtually no interest in adopting such strategies. Over the past several years some tentative and limited experiments with incentives have begun, most of them modifications of the marketable permits concept. Examples include EPA's "bubble" policy, emission offsets, and proposals for averaging automobile emission standards. None of these plans, however, is a full-blown incentive scheme of the type advocated by economists, and the overwhelming majority of environmental regulations continue to rely on traditional, uniform emission standards.

Much of the opposition to incentives has a strong emotional content; opponents sometimes characterize charges and permits as "licenses to pollute," although environmentalists who support incentives refer to the "pollutor pays principle."[6] Kneese and Schultze (1975) argue that reliance on direct regulation also reflects the legal training of most legislators; lawyers are used to thinking in terms of sharply defined rights and obligations, enforced by the courts, and not in terms of the economic incentives that economists find so natural. Schultze (1977) also discusses what he calls the "mystery of the market," the fact that few noneconomists have a very clear understanding of the role of prices in allocating scarce resources. Rather than rely on the indirect, somewhat mysterious effect of incentives, they prefer direct action through the imposition of standards.

Opponents of incentives also have leveled a variety of more substantive criticisms. Under the classical formulation an emission charge should be set equal to the marginal damage caused by emissions. That is, they should reflect the harm caused by incremental emissions and the dollar value ascribed to reducing that harm. In essence, the regulator must decide how much it is worth to eliminate a unit of emissions. With most pollutants, however, we are highly uncertain about the magnitude of the physical damages caused and have no well-accepted means for valuing those damages in monetary terms. In dealing with low-level environmental exposures to carcinogens such as benzene, for example, alternative risk estimates often span several orders of magnitude, and there is little agreement as to how much society should be willing to spend to "save a life." Under such circumstances, critics argue, damage-based charges can never be more than a theoretical toy.[7]

Critics also question the feasibility of monitoring and enforcing compliance with incentive-based schemes. Analyses of charges, particularly theoretical ones, typically assume that the charge can be levied directly on emissions. Often, however, it is extremely expensive, if not impossible, to measure emission levels accurately over time, in which case strategies based on direct monitoring are impracticable.

These criticisms have not gone unanswered. The problem of estimating the damages caused by emissions is a difficult one, but it applies to standards as well as to incentives; setting the appropriate standard requires knowledge of the damages and the ability to place at least rough bounds on their monetary value. Similarly, monitoring problems are by no means unique to incentives; emission standards also require an ability to measure emissions, and while direct monitoring may be the ideal, indirect methods can be used for charges or permits, just as they are used now for standards. Emissions may be sampled, for example, or estimated from information on control equipment and production levels.

Choosing the Right Target

The standards vs. incentives debate, by focusing on variations in the marginal costs of controlling emissions, has missed a key issue—variability in the marginal damages caused by emissions. The goal of environmental regulation is not to reduce emissions per se but rather to decrease damages. Thus regulatory strategies should be sensitive to intersource differences in marginal damages as well as to costs. The variability in damages is often at least as great as that in costs. In the case of benzene emissions from maleic anhydride plants, for example, the estimated damage caused by a unit of emissions varies by a factor of 50 across plants, while the cost differences appear to be much smaller. Just as it makes little sense to impose the same emission limit on plants with widely differing control costs, so too is it inefficient to impose a uniform limit on plants with radically different marginal damages. Uniform emission standards of the sort proposed for maleic anhydride plants fail to take account of either type of heterogeneity.

Traditional incentive schemes targeted on emissions do an excellent job of dealing with heterogeneity in costs but perform no better than uniform standards with respect to variations in benefits. They provide no incentive for plants at high-damage sites to engage in greater control efforts, nor do they encourage firms to choose low-damage sites. Thus we need to devise incentive schemes that are responsive to differences in marginal damages, that direct control expenditures where they will yield the highest returns. One approach is to vary the incentive for emission control with the level of marginal damage. Another is to alter the target, to apply the incentive to a measure more closely related to damages such as exposure.

These same alterations in targets also can be used to make standards more efficient. Tightening standards in urban areas and relaxing them in rural areas, for example, might yield substantial reductions in both costs and damages. In many cases such modifications in standards may yield greater gains than would switching to uniform, emission-targeted incentives.

We also need to consider alternative targets because emissions may be very difficult, or even impossible, to monitor. Benzene emissions from service stations, for example, cannot be measured continuously. We can, however, estimate them reasonably well if we know the benzene content of the gasoline, the efficiency of the vapor-recovery equipment (if any), and the volume of gasoline pumped. EPA considered imposing limits on the benzene content of gasoline and also standards requiring vapor-recovery devices. The incentive-based parallels might be, respectively, a charge on the benzene content of gasoline and a charge levied on service stations as determined by the volume of gasoline sold and the vapor-recovery devices in use. An alternative would be to use the same information to construct a single measure on which a charge would be levied.

Economists have paid little attention to such targeting issues, letting the choice of instruments—the charges vs. standards debate—hold center stage. Regulators, on the other hand, have devoted relatively little attention to the choice of instruments; both legislation and custom tell them to use standards. The targeting problem, in contrast, is of immediate practical concern; it in large part determines the ease with which regulations can be monitored and enforced. Neither the academics nor the practitioners, however, have recognized, much less analyzed systematically, the efficiency implications of alternative targets or the interactions between the choice of instruments and targets.

This book provides such an analysis. It examines the choices of both targets and instruments and shows how appropriately targeted economic incentives could yield large improvements in the efficiency of environmental regulation. Most of the analysis is applicable to the full range of environmental problems, though some of it is specific to airborne carcinogens, such as benzene. The general principles, however, in particular the importance of focusing on the marginal benefits of regulation as well as its marginal costs, have far broader applicability.

The primary emphasis throughout the book is on achieving economic efficiency, defined in the usual manner as maximizing net benefits. Much of the analysis is theoretical, but numerous empirical examples are used, in addition to a detailed case study. Even in the conceptual portions a major effort is made to give readers an idea of the magnitudes as well as the signs of the efficiency differences among alternative strategies. Although the focus is on efficiency, issues of equity and fairness are also addressed at several points.

The first three chapters analyze the choice of instruments, comparing standards, charges, permits, and subsidies under the traditional assumptions that emissions are the appropriate target of regulation. Chapters 5 through 7 provide the major original theoretical contributions, analyzing the interactions between alternative instruments, primarily charges and standards, and alternative targets, in particular emissions and exposure. Chapters 8 and 9 illustrate these concepts through a case study of benzene emissions from maleic anhydride plants. The final chapter summarizes the arguments and offers a concrete proposal for using exposure charges in a "generic" policy for airborne carcinogens.

Chapter 2

Alternative Instruments and the Allocation of Control Efforts

Regulators face a complex problem in designing efficient strategies for controlling environmental externalities. The range of issues that potentially deserve examination is enormous, as the voluminous literature on the subject attests. It is easier to understand the key issues, however, if we begin with a basic model and then build in additional complexity gradually, relaxing assumptions one at a time. The initial focus is on how well alternative instruments—charges, standards, permits, and subsidies—allocate emission control efforts across firms.

Consider a highly simplified version of the problem of environmental externalities. One or more firms emits a hazardous substance as the by-product of producing some good. Each firm produces a fixed level of output, but it can reduce its emission rate through expenditures on control efforts. It seeks to minimize its costs. The government has the authority to limit emissions through any of the four instruments listed here. Its goal is to minimize the social costs of production, which are the sum of the costs borne by the firms and the damages to others caused by the emissions. The government is financed by lump-sum taxes so that the implications of alternative schemes for government revenues and expenditures can be ignored to the extent that they are merely transfers. Moreover the regulatory authority operates in a world of certainty, with perfect knowledge of the costs and benefits of control, at least at aggregate levels. Though admittedly unrealistic, these assumptions provide a useful starting point for the analysis.

Marginal Conditions for Efficient Control

A strategy that minimizes the sum of control costs and residual damages must have two characteristics: (1) control efforts are allocated among firms so as to minimize the cost of achieving any given level of aggregate emissions, and (2) the overall level of control reflects an appropriate balancing of the aggregate costs and benefits of control.

Single Firm

Let us begin with the case of a single emitter. Here, obviously, only the second efficiency condition is relevant. The firm's emissions impose damages on external parties for which the firm is not liable. These damages

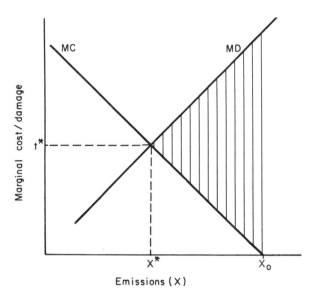

Figure 2.1   Optimal control with a single firm

might include adverse health effects, property damage, or such aesthetic losses as reduced visibility. At some cost the firm can reduce its emissions by modifying its production process or by installing emission control devices. In figure 2.1 the net marginal cost to the firm of controlling emissions is labeled MC, and is plotted as a function of the level of emissions. Note that the horizontal axis measures emissions, not emissions controlled, as is more common in such diagrams. The MC curve is drawn downward sloping on the usual assumption that marginal costs rise as the level of control increases; that is, as emissions are driven down the incremental cost of additional reductions increases. Marginal control costs often rise very sharply as emissions approach zero. Kneese and Schultze (1975), for example, report that it often costs as much to increase control from 97 to 99 percent as it does to control the initial 97 percent.

Figure 2.1 also includes a marginal damage curve, labeled MD, which represents the external damages imposed by the emissions. The MD curve is shown rising on the assumption that as emissions increase, each additional unit imposes higher incremental damages. Most theoretical analyses make this assumption of increasing marginal damage.[1] For certain types of pollutants, however, it seems more reasonable to assume that marginal damages are constant over the relevant range. If, for example, as many biomedical scientists believe, the risk posed by carcinogenic chemicals is proportional to the dose, then the marginal damage curve for such a substance would be a horizontal line.

In the absence of intervention the firm minimizes costs when it emits X_0 units, the point at which MC $= 0$ in figure 2.1. That level of emissions, however, is not optimal because marginal damages exceed marginal costs, and thus reductions in emissions would yield positive net benefits. In figure 2.1 the sum of control costs and damages is minimized at the emission level X^*, where MC $=$ MD. Beyond that point the marginal cost of control exceeds the marginal reduction in damage. The net efficiency gain due to reduced emissions is represented by the shaded triangle.

The regulator can achieve the optimum either directly, by imposing an emission standard at X^*, or indirectly, by charging the firm for emissions at the rate t^*, which corresponds to MC (and MD) at the optimum.[2] Faced with such a charge, the firm minimizes the sum of its control costs and its payments to the government at X^*. The sole difference between the two instruments is that under the charge the firm has to pay the government a total of t^*X^*, in addition to its control expenditures. That payment, however, is merely a transfer and thus does not enter into the evaluation of efficiency under the usual precepts of benefit-cost analysis. Alternatively, the regulator might offer the firm a subsidy for decreasing emissions. If, for example, the government paid the firm s dollars for each unit of emissions reduced from some base level, perhaps X_0, the firm would cut emissions to the level X^* if $s = t^*$; beyond that level the subsidy would not be sufficient to cover incremental control costs.

Multiple Firms
The choice between charges and standards becomes less trivial when the more complex and realistic case of multiple firms is considered. A simple model conveys the essential points and also introduces some basic notation that will be used in later chapters.

Suppose there are n plants to be regulated, where the level of emissions from the ith plant is denoted x_i. Let the cost of emission control for each firm be a function of its own emissions level:

$$c^i = C^i(x_i), \tag{2.1}$$

where the marginal cost of control (which is the negative of the marginal cost of additional emissions) is positive, $-C_x^i > 0$ ($C_x^i \equiv dC^i(x_i)/dx_i$), and increasing, $C_{xx}^i > 0$ ($C_{xx}^i \equiv d^2C^i(x_i)/dx_i^2$). Total cost is simply the sum of individual firms' costs:

$$C = \sum_{i=1}^{n} C^i(x_i). \tag{2.2}$$

Let total damages be a function of the unweighted sum of emissions from the different plants:

$$D = D\left(\sum_{i=1}^{n} x_i\right), \tag{2.3}$$

where marginal damages are positive and nondecreasing, $D' > 0, D'' \geq 0$. Note that this formulation implicitly assumes that for any given level of total emissions, marginal damage per unit of emissions is constant across plants. Though clearly unrealistic, this assumption is made in most of the theoretical literature on externalities.[3] Later, particularly in chapter 6, we shall analyze the impact of relaxing it.

Given these cost and damage functions, the optimization problem is to select the x_i's that minimize total social cost, S, the sum of equations (2.2) and (2.3). Differentiating with respect to each of the x_i's and setting the results equal to zero yields the well-known optimality conditions:

$$-C_x^i = D', \quad i = 1, \ldots, n. \tag{2.4}$$

At the optimum the marginal cost of control is constant across firms and is equal to marginal damage. In theory this result can be achieved by direct regulation, by imposing a separate standard on each firm. That is, the ith firm faces the constraint $x_i \leq x_i^*$. Alternatively, it can be achieved by levying a uniform charge on emissions at a rate equal to the marginal damage caused by a unit of emissions at the optimum, $t^* = D'(\sum_{i=1}^{n} x_i^*)$. Faced with that charge, the ith firm seeks to minimize the sum of its control costs and charge payments, $C^i(x_i) + t^* x_i$, which it achieves when

$$-C_x^i = t^* = D', \tag{2.5}$$

as in equation (2.4). The obvious advantage of the charge scheme is that a single charge rate can be set, rather than n different emission limits. Moreover regulators need not have cost data for each plant, as they must to set individually tailored standards; data on average marginal control costs are sufficient.

In practice, of course, separate standards are rarely set for individual emission sources but rather are uniform across large classes of plants. The regulator's problem is to select a uniform standard, $\bar{x}$, so as to minimize

$$S = \sum_{i=1}^{n} C^i(x_i) + D\left(\sum_{i=1}^{n} \bar{x}\right), \tag{2.6}$$

which yields the optimality condition

$$\frac{\sum_{i=1}^{n} C_x^i}{n} = D'(n\bar{x}^*). \tag{2.7}$$

That is, the standard should be set at the point where the average marginal control cost of the sources is equal to marginal damage. If marginal costs vary across firms, the total social cost with a uniform standard must be higher than with a uniform charge.

Consider the hypothetical case of two chemical plants located adjacent to one another, both of which emit benzene. Under a uniform standard both must meet the same limit on emissions. At that limit the marginal cost of controlling benzene emissions is $1 per kilogram at plant A but only $.20 per kilogram at plant B. The result is clearly inefficient. Suppose we allowed plant A to increase its emissions by 1,000 kg, thus saving $1,000, while at the same time we required plant B to reduce its emissions by 1,000 kg, at an additional cost of $200. The net result would be a savings of $800, with no change in overall emissions or external damages. Alternatively, we could let plant A reduce its control expenditures by $1,000, thus generating an extra 1,000 kg in emissions, while requiring plant B to increase its expenditures by $1,000, thus reducing its emissions by 5,000 kg. The net result would be to lower emissions by 4,000 kg at no increase in total cost.

This simple example illustrates the inefficiency of standards relative to charges; when marginal control costs vary across firms, it is possible to decrease either costs or damages while holding the other constant. For any given level of total emissions control, costs will be higher under a uniform standard than under a charge scheme. Thus total social costs, the sum of the costs of control and residual damages, will be lower with the optimal charge than with the optimal standard.

Optimal Control Level
The reader should note that the optimal level of total emissions will, in general, depend on the regulatory instrument. If the functions are "well behaved," the optimal level of total emissions under either a charge or a standard will be the point where marginal benefit equals aggregate marginal cost. The marginal benefit curve will be the same for both instruments. The marginal cost curves, however, almost certainly will differ because the allocations of control efforts will be different. The alternative regulatory instruments are analogous to alternative control technologies,

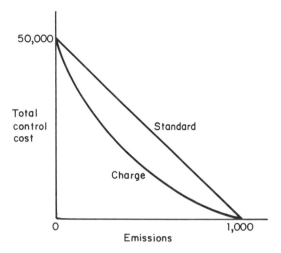

Figure 2.2   Total control costs

with different cost functions, and just as the optimal level of control will depend on the costs of the technology available, so too will it vary with the instrument chosen.

Intuitively, one might assume that the aggregate marginal cost curve under a charge scheme would always be lower than that under a standard, because total costs are always lower. Based on that assumption, the optimal level of control always will be tighter with a charge than with a standard. There are many cases, however, where marginal cost with a charge is higher over some levels of control. Consider the following example. A very large number of equal-size sources currently emit a total of 1,000 units of some hazardous substance. Each source can control its own emissions at constant unit cost, but that cost varies across sources. In particular, the distribution of marginal costs across sources is uniform over the range $0 to $100.

Under a charge scheme the aggregate marginal cost curve rises from $0 to $100 as the level of total emissions falls from 1,000 to 0. Under a uniform standard, however, aggregate marginal cost remains constant at $50, because at any level of emissions the marginal control burden is spread evenly across firms. Thus for the first 500 units of control aggregate marginal cost is lower under the charge scheme, but for the second 500 units it is lower for the standard. (Over the whole range, of course, total costs are lower with the charge.) As a result the optimal level of control with a charge may be higher or lower than with a standard depending on the location of the marginal damage function.

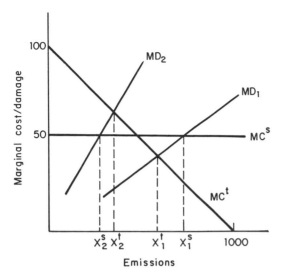

Figure 2.3   Optimal control levels of alternative instruments

Figures 2.2 and 2.3 illustrate this example. Figure 2.2 plots total control costs under each instrument, showing that total cost is lower with the charge then with the standard at every emission level other than 0 and 1,000. In figure 2.3 the horizontal line labeled MC^s is the aggregate marginal cost curve under a standard, whereas MC^t is the corresponding curve under a charge. If the marginal damage curve is MD_1, then the optimal level of emissions is X_1^s under the standard and X_1^t under the charge, where $X_1^t < X_1^s$ (i.e., the charge leads to tighter control). If, however, the marginal damage curve is higher, say, MD_2, then the optimal level of emissions under the standard (X_2^s) is lower than under the charge (X_2^t).

This simple example also illustrates two more general propositions. First, the absolute gains of shifting from a uniform standard to a charge are likely to be greatest at intermediate levels of control. At low levels of control, control costs are low, even with the standard, so the potential gains are limited. To achieve very high levels of control, most control options must be exercised for most sources, so little room exists to reallocate control efforts among sources when the switch is made from the standard to the charge.

The second proposition is that the tighter the level of aggregate control, the more likely it is that aggregate marginal cost will be higher with a charge than with a standard because the charge will have used up the low-cost options at lower levels of control, while the standard will have failed to exercise some of those low-cost options earlier. Thus we might expect that the higher the level of marginal damage, the more likely that the optimal charge would result in a higher level of emissions than the optimal uniform standard. For advocates of strict environmental regulation that may sound like a drawback for charges. It is critical to remember, however, that the charge always can be set to yield the same level of emissions as the optimal standard. In that case total control costs will be lower than with the standard (and, by construction, total damages will be the same), but total social cost will be higher than with the optimal charge.

Permits and Subsidies
In the highly simplified context examined thus far, two additional incentive-based approaches, subsidies and marketable permits, yield the same results as an emission charge. Under an emission-reduction subsidy, as explained earlier in the case of a single firm, each firm receives a subsidy for reductions in emissions from some base level, x_i^B. That benchmark might, but need not, be the firm's initial emission level. Under such a scheme each firm seeks to minimize control costs minus the subsidy, $C^i(x_i) - s(x_i^B - x_i)$. At the firm's optimum the marginal cost of control is equal to the subsidy rate:

$$-C_x^i = s. \tag{2.8}$$

Thus, as a comparison of equations (2.5) and (2.8) reveals, a subsidy at the rate s yields the same results as a charge levied at that rate. The result should hardly be surprising, as the subsidy scheme can be thought of as a lump-sum transfer to the firm of sx_i^B, coupled with a charge per unit of emissions at the rate s.

In contrast to subsidy and charge schemes, which set a price and allow the market to set the total level of emissions, marketable permit approaches set a total quantity of emissions and let the market set the price.[4] Under a marketable permits approach the government issues a number of permits equal to the total desired level of emissions. These permits may be allocated initially in a variety of ways. The regulatory authority, for example, might auction them to the highest bidders, or it might distribute them to existing firms on some other basis such as current emissions. Once the permits are allocated, however, firms are free to buy and sell them as they wish. In a static context with no uncertainty, the ultimate distribution of emissions will be the same as under the corresponding charge scheme. That is, if a charge rate of t^* leads to a total of X^* units of emissions, then if X^* permits are issued their equilibrium price should be t^*, assuming a competitive market in the permits and no transactions costs. As shown in chapter 4, however, permit and charge schemes differ in their efficiencies under conditions of uncertainty and over time, as costs and damages change.

Estimated Gains from Incentives

The marginal conditions derived in the previous section suggest that switching from uniform standards to charges, or to some other incentive-based instrument, would yield efficiency gains. The pure theoretician may find comparisons of marginal conditions sufficient, but regulators, legislators, and even policy-oriented economists, quite reasonably, are likely to want some indication that the gains will be significant before they will be inclined to adopt a very different mode of regulation.

The magnitude of the gains obviously will vary across different situations, depending on a variety of factors. In some cases the gains may be negligible, in others quite large. Chapter 9 compares the net benefits of alternative regulatory approaches for a particular pollutant and a particular industry, benzene emissions from maleic anhydride plants. At this point, however, we want a rough idea of the range of empirical estimates and an indication of the variables that help determine the size of the gains.

Empirical Estimates

The efficiency gain ideally should be measured as the difference in net benefits (reduction in damage minus control costs) under an optimal charge scheme as opposed to a uniform standard set at the optimal level. That approach, however, requires placing a dollar value on damages. To avoid that sensitive and difficult problem, most studies instead have compared the costs of achieving a given ambient standard using different regulatory instruments. The limited empirical results available indicate that the cost savings may be quite impressive. A study of the Delaware Estuary, for example, estimated that a simple uniform effluent charge could achieve a dissolved oxygen level of 2 parts per million (ppm) at an annual cost of $2.4 million, compared to a cost of $5.0 million under a standard requiring uniform treatment of wastes, a savings of over 50 percent.[5] For a more ambitious water-quality target of 3 to 4 ppm, the charge would impose costs of $12.0 million annually, as compared to $20.0 million under the standard, a savings of about 40 percent. Atkinson and Lewis (1974) found still larger proportional differences for controlling airborne particulates in St. Louis. Compared to least-cost emission control programs of the type achieved by emission charges, Atkinson and Lewis estimate that standards requiring uniform percentage reductions in emissions would cost six times as much to achieve the primary standard and one and a third as much to achieve the secondary (tighter) standard.

A Theoretical Model

The empirical estimates cited here, together with those derived from other similar studies, suggest that the cost savings to be realized in shifting from direct regulation to incentive-based schemes are likely to be significant in many instances. These empirical results, however, do not provide any insight into the factors that influence the magnitudes of the savings in different cases. The simple model developed here helps both to highlight some of the critical factors and to establish some basic results for comparison with more complex cases examined in subsequent chapters. Although the explicit comparison is between standards and charges, the reader should keep in mind that the results for charges hold as well for marketable permits and emission-reduction subsidies.

To obtain expressions for the relative costs of charges and standards, we must specify the functional forms of the cost and damage functions and the distribution of costs across emission sources. Although the use of specific functional forms limits the generality of the results, it is necessary if we are to go beyond comparisons of marginal conditions.[6] The results derived here are of limited intrinsic interest; their primary purpose is to establish some basic results for comparison with analyses in later chapters that employ closely related models.

Let the cost of controlling emissions from the ith unit of production be given by

$$C^i(x_i) = a_i x_i^{-\alpha}, \quad \alpha > 0, \tag{2.9}$$

where α is constant across sources but the coefficient a_i varies. The marginal cost of control is

$$-C^i_x(x_i) = \alpha a_i x_i^{-\alpha-1}. \tag{2.10}$$

Note that, in keeping with most "real" control-cost functions, this one exhibits increasing marginal costs as the level of control increases (level of emissions decreases). The exponent α may be interpreted as the elasticity of control costs with respect to emissions, and by varying α, we can vary the curvature of the cost functions; the larger the value of α, the sharper the curvature and the more rapidly marginal costs rise as the level of emissions falls. If, for example, $\alpha = 1$, control costs double every time emissions are halved, whereas if $\alpha = 0.5$, costs double only when emissions are reduced by a factor of four. This functional form also has the convenient mathematical properties that the marginal cost of control at each source is positive at all levels of emissions and grows without bound as emissions (x_i) approach zero, thus ensuring that the first-order conditions will be sufficient for an optimum and that the complications of corner solutions will be avoided.[7]

On the damage side, let total damages be proportional to the total level of emissions:

$$D(x_1, \ldots, x_n) = \lambda \left(\sum_{i=1}^{n} x_i \right),$$ \hfill (2.11)

where λ is the shadow price of emissions, the product of the physical damages caused by each unit of emissions, and the dollar value placed on those damages. Assuming that exposure is proportional to emissions, this functional form is consistent with the widely used linear dose-response model of carcinogenesis. Results also have been derived for a more general damage function of the form $D(x_1, \ldots, x_n) = \lambda(\sum_{i=1}^{n} x_i)^\beta$, which, for $\beta > 1$, exhibits increasing marginal damages. In the interest of brevity those results are not reported here in any detail. Qualitatively, they are the same as for the linear damage function ($\beta = 1$). Quantitatively, increases in β magnify the proportional differences.[8]

Given these cost and damage functions, total social cost is

$$S = \sum_{i=1}^{n} a_i x_i^{-\alpha} + \lambda \sum_{i=1}^{n} x_i,$$ \hfill (2.12)

where the first summation is the total cost of control and the second is the damage due to emissions. Differentiating with respect to each x_i, we obtain the emissions levels that minimize social costs:

$$x_i^* = \left(\frac{\alpha a_i}{\lambda} \right)^{\varepsilon}, \quad i = 1, \ldots, n,$$ \hfill (2.13)

where $\varepsilon \equiv 1/(\alpha + 1)$ is introduced to simplify this and subsequent expressions. Substituting equation (2.13) for each x_i in equation (2.12) yields an expression for minimum social cost:

$$S^* = \left(\frac{\lambda}{\alpha} \right)^{\alpha\varepsilon} (1 + \alpha) \left(\sum_{i=1}^{n} a_i^\varepsilon \right) = \left[\left(\frac{\lambda}{\alpha} \right)^{\alpha\varepsilon} (1 + \alpha) n \right] E(a^\varepsilon),$$ \hfill (2.14)

where

$$E(a^\varepsilon) \equiv \frac{\sum\limits_{i=1}^{n} a_i^\varepsilon}{n}.$$

As expected, this same result can be achieved by imposing a uniform charge on emissions at the rate λ. If the charge rate is t, the ith firm seeks to minimize the sum of control costs and charge payments, $a_i x_i^{-\alpha} + t x_i$, which yields

$$x_i^* = \left(\frac{\alpha a_i}{t}\right)^\varepsilon. \tag{2.15}$$

If $t = \lambda$, equation (2.15) is identical to equation (2.13), so equation (2.14) also represents total social cost under the optimal charge. If the charge is viewed as the price of emissions, ε may be interpreted as (the absolute value of) the firms' own-price elasticity of demand for the right to emit.

Under a uniform emission standard, $\bar{x}$, $x_i = \bar{x}$ for all sources. Thus social cost is given by

$$S_s = \sum_{i=1}^{n} a_i \bar{x}^{-\alpha} + \lambda \sum_{i=1}^{n} \bar{x}. \tag{2.16}$$

Differentiation yields the optimal standard

$$\bar{x}^* = \frac{\alpha \sum\limits_{i=1}^{n} a_i^\varepsilon}{\lambda n}, \tag{2.17}$$

which is the emission level at which average marginal cost is equal to marginal damage. Substituting equation (2.17) in equation (2.16), we obtain an expression for total social cost under the optimal standard:

$$S_s^* = \left(\frac{\lambda}{\alpha}\right)^{\alpha\varepsilon} (1 + \alpha) \left(\sum_{i=1}^{n} a_i\right)^\varepsilon = \left[\left(\frac{\lambda}{\alpha}\right)^{\alpha\varepsilon} (1 + \alpha) n\right] [E(a)]^\varepsilon. \tag{2.18}$$

Note that the only difference between equations (2.18) and (2.14) is that in equation (2.18) the expectation of a is taken before it is raised to the power ε. Since $\varepsilon \equiv 1/(\alpha + 1) < 1$ for $\alpha > 0$, it is clear that the social cost under the standard must be higher than that under the charge as long as there is any variation in a_i across firms.

The difference in social costs under the two regulatory regimes is simply the difference between equations (2.18) and (2.14):

$$S_s^* - S^* = \left[\left(\frac{\lambda}{\alpha}\right)^{\alpha\varepsilon}(1+\alpha)n\right]\{[E(a)]^\varepsilon - E(a^\varepsilon)\}. \tag{2.19}$$

Note that this approach differs from the usual one in that it does not compare the control costs of achieving the same level of emissions, but rather recognizes that the optimal level of total emissions will depend on the regulatory approach used. With these particular cost and damage functions the charge yields both lower costs and lower emissions. We can put the savings in social cost under the standard in relative terms by dividing equation (2.19) by control costs under the optimal standard. As control costs under each alternative comprise the fraction $\varepsilon = 1/(\alpha + 1)$ of total social cost, the result is

$$\begin{aligned}\Delta S^* &= \frac{S_s^* - S^*}{\varepsilon S_s^*} = (1+\alpha)\left(1 - \frac{S^*}{S_s^*}\right) \\ &= (1+\alpha)\left\{1 - \frac{E(a^\varepsilon)}{[E(a)]^\varepsilon}\right\}.\end{aligned} \tag{2.20}$$

Distribution of Costs
To proceed further, we need to specify the distribution of the cost coefficient, a, across firms. Intuitively, it is clear that this distribution will be of central importance. If costs vary only slightly, little is lost by treating all firms as if they were the same, as the uniform standard does. If costs vary widely, however, a uniform standard is likely to be extremely costly. In practice, of course, the distribution of costs is discrete, because the number of sources affected by a regulation is finite. For analytic purposes, however, it is more convenient to work with continuous distributions. A variety of different distributions might be hypothesized. Unfortunately, the expression for the relative difference in social costs derived in equation (2.20) includes the expectation of a^ε, where ε is not an integer, which means that the use of most probability distributions would necessitate numerical simulation. Moreover some of the models derived in later chapters include the expectations of the products of more than one variable, so that it is desirable that the probability distribution chosen here be generalizable to multiple variables.

The log-normal distribution is an attractive candidate both because it yields closed-form analytic expressions and because it is easily extended to the multivariate case. It is also skewed to the right, which matches, at least qualitatively, many actual distributions of control costs (i.e., there is a thin "tail" of high control-cost firms). Leone and Jackson (1978), for example, estimated the costs per ton of achieving "best practicable technology" (BTP) water effluent standards in the tissue portion of the pulp and paper industry. They found a wide distribution of costs, from $1.85 to $82.82 per ton of tissue, that was highly skewed, with 80 percent of the firms having costs below $12.34 per ton and an overall mean of $9.41 per ton.

If we assume that a is distributed log-normally, the required expectations can be obtained using the moment-generating function of the normal distribution. The moment-generating function, $M(v)$, for a probability density distribution, $f(z)$, is defined as

$$M(v) = E(e^{vz}),$$

where v is any real number. If z is distributed normally with mean μ_z and variance σ_z^2,

$$M(v) = \exp\left(\frac{v\mu_z + v^2\sigma_z^2}{2}\right),$$

where $\exp(z) \equiv e^z$.[9] If $z = \ln Z$, then $M(v) = E(Z^v)$, which is precisely the kind of expectation for which we need to find an expression. Thus, if we assume that the natural log of a is distributed normally, with mean μ_a and variance σ_a^2,

$$E(a^\varepsilon) = \exp\left(\frac{\varepsilon\mu_a + \varepsilon^2\sigma_a^2}{2}\right), \tag{2.21}$$

and

$$[E(a)]^\varepsilon = \exp\left(\frac{\varepsilon\mu_a + \varepsilon\sigma_a^2}{2}\right). \tag{2.22}$$

Using equations (2.21) and (2.22), the relative savings in social cost under the charge, equation (2.20), may be rewritten:

$$\Delta S^* = (1+\alpha)[1 - \exp(-0.5\alpha\varepsilon^2\sigma_a^2)]. \tag{2.23}$$

Note that, because $\varepsilon \equiv 1/(\alpha + 1)$, the value of this expression depends on only two parameters, α (the elasticity of control costs with respect to emissions) and σ_a^2 (the variance of the log of the cost coefficient). It is easily shown that an increase in the value of either parameter increases the relative advantage of the charge. That is, the more rapidly marginal

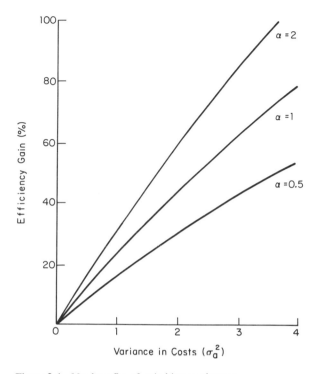

Figure 2.4   Net benefits of switching to charges

costs rise with the level of control and the greater the dispersion of costs across firms, the greater the proportional savings in shifting from a standard to a charge.

Figure 2.4 plots the relative savings as a function of σ_a^2 for several values of α. As the graph makes clear, both variables are important in determining the level of savings. If $\alpha = 1$, for example, the savings range from 24 percent for $\sigma_a^2 = 1$ to 79 percent for $\sigma_a^2 = 4$.[10] Holding constant $\sigma_a^2 = 2$ and varying α, the savings range from 30 percent for $\alpha = 0.5$ to 60 percent for $\alpha = 2$. Although the selection of any particular pair of parameter values is arbitrary, it is useful to select a base case for comparison in future models, where additional parameters are introduced. Let the base values be $\sigma_a^2 = 2$ and $\alpha = 1$ ($\varepsilon = 0.5$), which yield a savings of 44.2 percent, roughly consistent with the empirical estimates cited earlier, though perhaps a bit on the low side. With the same value of σ_a^2, but $\alpha = 0.5$, the savings is 29.9 percent.[11] Appropriately structured permits or subsidies yield these same efficiency gains.

Impacts on Firms

Intuition might suggest that the efficiency gains under charges would lead to a reduction in costs for most firms. It is important to remember, however, that the efficiency gains may include reductions in damages—which represent a benefit to society but not to the firms subject to regulation—and that the charge payments, although a transfer from a societal perspective, are a cost to firms. Thus it is likely that most firms will find their costs associated with regulation higher under charges than under standards.

Consider a "typical" firm, one for which the optimal uniform standard is also the optimal level for that particular plant. Under a charge its emission-control expenditures will not change, but it will have to pay for the damage its remaining emissions impose, so its total costs will rise. Firms with lower than average control costs also will face higher total costs; the charge will lead them to engage in tighter control, and they will have to pay for residual damages. Only firms with much higher than average control costs will find total costs reduced, as the savings from reduced control more than offset the increase due to the payment for uncontrolled emissions.

Figure 2.5 illustrates the argument. Marginal damage is given by the MD curve, and MC_0 is the marginal cost for the average firm. The optimal uniform standard is $\bar{x}^*$. If we switch to the optimal charge, $t^* = MD$, the typical firm continues to control at the same level, so its costs rise by the charge on remaining emissions, $t^*\bar{x}^*$. Other firms, however, adjust their emission levels to reduce the net increase in costs. A low-cost firm, with marginal costs given by MC_L in figure 2.5, for example, cuts its emissions to x_L^*, thus reducing the increase in total costs by the shaded area to the left of the base level of emissions. A high-cost firm, MC_H in the figure, lowers its net increase by the shaded area to the right of $\bar{x}^*$ by increasing its emissions; if the change is large enough, its costs will be lower with the charge than with the standard.

Model Results

We can investigate this issue further using the model developed previously. Let $\bar{F}_s$ be the cost for the typical firm of complying with the optimal standard, $\bar{x}^*$, as given in equation (2.17). This firm's cost coefficient is $a = E(a)$. With a charge of $t^* = \lambda$, the firm continues to emit the same amount, but it must pay residual damages. Under this model damages are equal to α times control costs. Thus the cost to the typical firm under

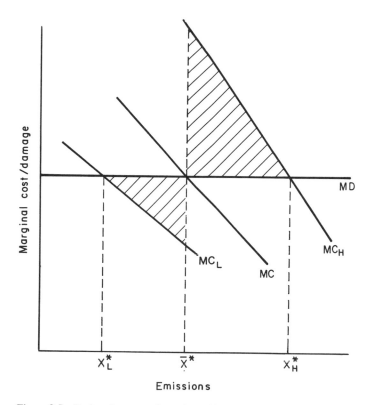

Figure 2.5   Reduced costs to firms that adjust control levels

the charge will be

$$\bar{F}_t = (1 + \alpha)\bar{F}_s. \tag{2.24}$$

If, for example, $\alpha = 1$, the typical firm's costs under the charge will be double what they are under the standard. If $\alpha = 0.5$, they will be 50 percent higher than under the standard.

Total costs across all firms, however, do not increase by as large a factor because firms with higher or lower cost coefficients are able to reduce their net increases by altering emission levels. The total cost to firms under the charge is the same as total social cost given in equation (2.14). Under the standard firms bear only the control costs, which comprise the fraction ε of total social cost in equation (2.18). Thus the ratio of firms' costs under the charge to that under the standard is given by

$$\frac{S^*}{(\varepsilon S_s^*)} = \frac{(1 + \alpha)E(a^\varepsilon)}{[E(a)]^\varepsilon}. \qquad (2.25)$$

If we again assume a log-normal distribution for a, equation (2.25) may be rewritten as

$$\frac{S^*}{(\varepsilon S_s^*)} = (1 + \alpha)\exp{(-0.5\alpha\varepsilon^2\sigma_a^2)}. \qquad (2.26)$$

For our base values of $\sigma_a^2 = 2$ and $\alpha = 1$, this expression takes on the value 1.56; total costs to firms, including charge payments, are 56 percent higher with the charge than with the standard. For $\alpha = 0.5$, the increase is only 20 percent, primarily because control costs comprise two-thirds of total social cost.

These results suggest an important reason for the lack of enthusiasm shown by most firms for the use of charges rather than standards: although social costs would fall, costs to firms would rise in most cases. Moreover the sharpest increases would be borne by firms with average control costs; charges would provide a *relative* advantage to firms with higher- or lower-than-average costs.

Permits allocated by auction will have exactly the same impact on firms' costs as charges. If the permits are given away, however, overall costs for firms are likely to be lower than with standards. (The gain to firms of course is offset by a reduction in government revenues from the sale of permits.) Using the same model as before, we can derive an expression for the ratio of total costs to firms under free permits to those under a uniform standard. Under both schemes firms as a whole do not pay for residual damages; they only pay the fraction ε of total social costs in equations (2.14) and (2.18). Thus the ratio is

$$\frac{\varepsilon S^*}{\varepsilon S_s^*} = \frac{E(a^\varepsilon)}{[E(a)]^\varepsilon}$$
$$= \exp{(-0.5\alpha\varepsilon^2\sigma_a^2)}, \qquad (2.27)$$

if the cost coefficients are distributed log-normally. For $\alpha = 1$ and $\sigma_a^2 = 2$, the ratio is 0.78; total costs to firms fall 22 percent as a result of the more efficient allocation of control efforts, despite the fact that overall control is tighter. The tightening of control, however, means that not all firms will reduce costs in the switch from a standard to free permits. Firms with average control costs will find themselves controlling at the same level but needing to purchase some additional permits from other firms. The

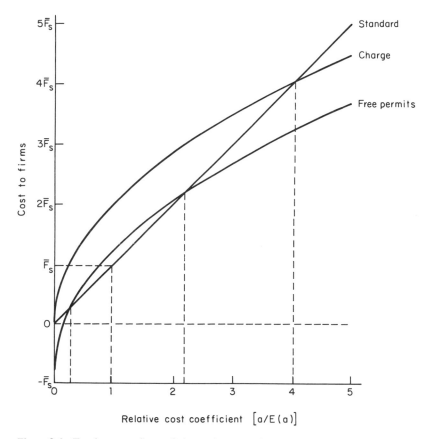

Figure 2.6 Total costs to firms of alternative strategies

gainers, both in absolute and relative terms, will be those firms with control costs substantially higher or lower than average.

Figure 2.6 plots the relative costs to firms as functions of the cost coefficient for the alternative regulatory strategies: a uniform standard, a charge (or permits allocated by auction), and free permits (where each firm initially is given the same number of permits per unit of capacity). The results are based on the model used here and a cost elasticity of $\alpha = 1$. Note that costs under the charge (or auctioned permits) are higher than those under the standard for all firms except those with cost coefficients four or more times higher than the average. Even with free permits firms face higher costs than with the standard unless their cost coefficients deviate significantly from the average. For a few firms with very low control costs sales of excess permits to other firms exceed the cost of control, so net costs are lower than without any regulation.

The only incentive scheme under which all firms are likely to reduce their net costs is the subsidy. Indeed with a subsidy all firms will find their costs lower than they would be in the absence of any intervention; unless the subsidy is higher than control costs, firms will not participate.

It is tempting to conclude from these results that subsidies or free permits are the most attractive options. Under the restrictive assumptions of this chapter, they yield the same efficient allocation of control efforts as a charge or permits sold at auction but do not engender the opposition of the regulated industry by increasing the costs it bears under a uniform standard. As shown in the next chapter, however, relaxing some of those assumptions reveals serious inefficiencies in both free permits and subsidies, particularly the latter.

Chapter 3

Refining the Static Analysis: Prices and Revenues

Early analyses of externalities focused on the problem that the prices of products do not reflect the external costs associated with their production or use. Most implicitly assumed that the externality and the final good were produced in fixed proportions, so the only way to reduce the externality was to reduce output of the good. The possibility of altering these proportions was, for the most part, ignored (Plott 1966). The classic economic solution was to impose a tax on the good equal to its external damages. More recently, analysts have tended to make the opposite set of assumptions, as we did in chapter 2; output of the final good is assumed to be fixed, and the reduction in damages occurs entirely through reductions in emissions per unit of output. If we think of emissions as a factor of production, the earlier analyses focused on the output or scale effect, whereas our analysis in chapter 2 addressed the substitution effect.

Neither polar set of assumptions is correct, though the fixed-output case is probably closer to reality in most instances. Substantial reductions in automobile emissions, for example, have been achieved by modifying engines and adding control devices rather than by reducing the output of automobiles, though higher auto prices due to emission-control requirements no doubt have had some minor effect on the number of automobiles. Nonetheless, the classic principle that the price of the product should reflect its full social cost, including external damages, remains valid.[1]

Impact on Final Product Prices

Let us begin our analysis of the impacts of alternative instruments on final product prices with the simple case of a competitive industry composed of identical firms with constant unit costs of production and constant marginal damages from emissions. In the absence of intervention the price will be equal to the marginal cost of production, which, given our assumptions, also will be the average cost. In figure 3.1 that price is labeled P_0. Each unit produced, however, generates emissions that cause external damages of D_0. Under the classical, output-oriented formulation, with damages per unit of output fixed, the appropriate intervention is a tax on the final good at the rate D_0 per unit. That yields an effective price to consumers of $P_1 = P_0 + D_0$, which decreases production of the final good from Q_0 to Q_1, thus reducing damages by $(Q_0 - Q_1)D_0$.

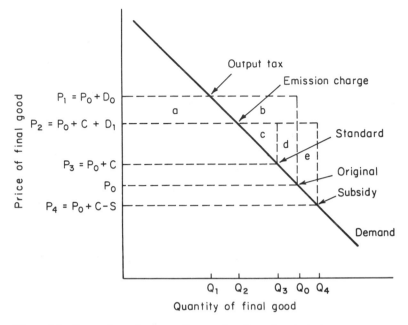

Figure 3.1   Comparison of output effects under alternative strategies

Subtracting the loss in consumer's surplus not captured in tax revenues leaves the net gain shown by the triangle $b + c + d$ in figure 3.1.

Consider now the possibility that damage per unit of the final good can be reduced with the addition of controls. Suppose that damages can be reduced from D_0 to D_1 at a cost of C, where $C + D_1 < D_0$; the controls decrease the social cost of producing the final good. This reduction can be achieved with any of the four instruments discussed in chapter 2 because, by assumption, all firms are identical, and thus no inefficiency is incurred by imposing a uniform standard. If we hold constant the quantity of output, all four instruments yield net benefits equal to the area of the rectangle $a + b$, which represents the reduction in social cost per unit $(D_0 - D_1 - C)$ times the number of units produced (Q_0).

The instruments differ in their output effects, however, and thus also in their total net benefits. Under the charge the firms' unit costs rise due to both the costs of control (C) and the charge on residual damages (D_1). In a competitive market price equals unit cost, so the price of the product rises to full social cost, $P_2 = P_0 + C + D_1$. Quantity produced falls to Q_2, yielding an additional net gain shown by the triangle $c + d$ in figure 3.1; that represents the reduction in external damages associated with

Table 3.1   Summary of net benefits for
figure 3.1

Regulatory strategy	Net benefit
Product tax	$b + c + d$
Emission charge	$a + b + c + d$
Emission standard	$a + b + d$
Emission subsidy	$a + b - e$

Note: Net benefit figures refer to areas in
figure 3.1.

the cutback in output, less the loss in consumer's surplus not captured by the government in charge revenues. Marketable permits yield the same result because from the firms' perspective the price of the rights is equivalent to a charge on emissions. Even if the government initially allocates the permits free of charge, they will have an opportunity cost because they could be sold to other firms.

Under the emission standard firms bear the cost of control but not the cost of the remaining damages. Thus the product price rises only to $P_3 = P_0 + C$, and the quantity falls by a lesser amount than with the charge or permits, to Q_3. Relative to the emission charge and permits, net benefits are lower by the area of the small triangle c.

The emission-reduction subsidy leads to a still lower price, as the cost of control is more than offset by the subsidy payments. (If the subsidy were not at least as great as the cost, firms would be better off maintaining their original emission levels.) Thus the product price under the subsidy— $P_4 = P_0 + C - S$, where S is the net subsidy per unit of output—is lower than the initial, preintervention price. The benefit from lowering the social cost per unit of output is reduced by the increase in the number of units produced. Relative to the emission standard, net benefits decrease by the area of the trapezoid $d + e$, which represents the additional external damages, plus the cost of the subsidy, less the gain in consumer's surplus. Table 3.1 summarizes the result for each instrument.

This analysis assumes that the base level of emissions from which reductions are subsidized depends on production levels. If, for example, such a subsidy were employed to reduce sulfur emissions from coal-fired electricity-generating plants, some baseline level of emissions per kilowatt generated from a typical uncontrolled plant might be established. Subsidy payments to each plant would then be the subsidy rate times the difference between the base rate and the achieved rate, times the number of kilowatts generated. The effect of the subsidy would be to reduce the cost of electricity, with the results described here.

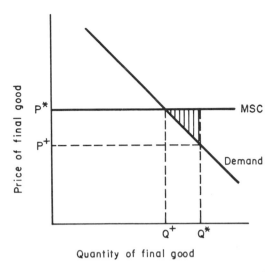

Figure 3.2 Efficiency loss when price below marginal social cost

In theory the problems associated with this type of subsidy could be eliminated by paying the subsidy for potential as well as actual production. That is, the baseline could be made independent of actual production. In practice, of course, it is impossible to do so because the regulator would have to estimate potential production not only for existing plants but also for possible new entrants.[2] Efficient subsidy schemes can be devised, but in essence they collapse to a charge with lump-sum compensation.

The key problem is that both a subsidy and, to a lesser extent, an emission standard lead to the product price being below marginal social cost. The shaded area in figure 3.2 shows the welfare loss due to pricing below marginal social cost (MSC). The optimal price is $P^* = MSC$, whereas the suboptimal price is P^+; Q^* and Q^+ are the corresponding quantities. The welfare loss is given by

$$0.5(P^* - P^+)(Q^+ - Q^*) = 0.5\eta\left(\frac{P^* - P^+}{P^*}\right)^2 P^*Q^*, \tag{3.1}$$

where η is the own-price elasticity of demand for the final good at the optimal price. The right-hand side of equation (3.1) shows that the welfare loss is proportional to the own-price elasticity of demand and to the square of the proportional deviation of price from marginal cost. Under a standard the divergence between price and marginal social cost will be equal to residual damage. Thus the larger the level of residual damage relative to production and control costs, and the larger the elasticity of demand, the less attractive is the standard as compared to the charge or permits. Under a subsidy the divergence between price and marginal social cost is even greater. If the divergence is great enough and and the elasticity of demand is high enough, a subsidy scheme actually can yield negative net benefits; under such conditions the welfare losses due to increased production outweigh the reduction in external damages per unit produced.

Heterogeneous Firms
The analysis becomes more complicated and less amenable to formal analysis when we drop the assumption that firms are identical, with constant and uniform unit costs. Given the inefficiencies of uniform standards in allocating control efforts, we can no longer be certain that the equilibrium price of the product will be lower with a standard then with a charge or marketable permits. In such a case the marginal private cost of the marginal firm may be higher with a standard than with a charge, even though the latter will include the residual damage from emissions. Thus the standard may lead to a higher price and a lower level of production than the charge. Despite this fact production under the standard will still be excessive in the sense that it will be higher than it would be if the price included the damages associated with production of the marginal unit. The subsidy will be inefficient in the same sense; price will be less than marginal social cost.

Standards often do take account of one important source of heterogeneity in costs: standards typically are tighter for new sources than for existing ones. The efficiency-based rationale for such distinctions is clear; it usually costs more to add controls to existing plants than to include them in new construction. Thus under any of the incentive schemes old plants would control less on average than new ones. Firms would have to pay for those higher emissions, however—in the form of higher charge payments, the purchase of additional permits, or the receipt of lower subsidies—so they would have the appropriate incentive to replace old plants with new ones. In contrast, with differential standards firms do not bear the external costs imposed by higher emissions from older plants and thus do not have a sufficient incentive to replace them. Indeed, if the differential is high enough, they may continue to operate existing plants longer than they would have had there been no regulation at all. Similar problems confront attempts to make standards more lenient for smaller firms, which often have higher marginal control costs.

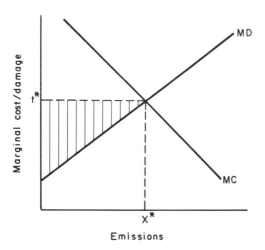

Figure 3.3   Total charges in excess of total damages when marginal damages increasing

Increasing Marginal Damages

Rose-Ackerman (1974) argues that a charge set equal to marginal damage will impose excessive costs on firms if marginal damages are increasing. The basis for her argument is straightforward: firms are charged at the same rate for all emissions, although only the last unit of emissions does as much damage as the charge would indicate. Thus total payments by the firms will exceed total damages, driving some firms out of business that would survive if the sum of the charge payments equaled total damage. Rose-Ackerman contends that the results are inefficiently small numbers of firms and low levels of output. Figure 3.3 illustrates the argument. To achieve the emissions level X^*, the charge is set at t^*. Total payments by firms are t^*X^*, but total damage is only the area under the MD curve from $X = 0$ to $X = X^*$, so the shaded area represents the excess of payments over damages.

Rose-Ackerman's argument is superficially plausible, but misleading unless the number of firms affected is small. If the number of firms is large, so that a firm's emissions are small relative to the total, then in essence all of the emissions from that firm do damage at the marginal rate. Thus total damages equal total charges for each firm, though not for the industry as a whole. This result may seem counterintuitive, but consider the following example. Suppose that there are 1,000 identical firms, each releasing 10 units of emissions at the optimal level of control. Marginal damage is equal to $0.01X$, where X is the total level of emissions.

Thus at $X^* = 10,000$, marginal damage is 100, and the charge is 100 per unit. Each firm pays $10(100) = 1,000$ in total charges, and all firms together pay $1,000(1,000) = 1,000,000$. Actual damages, however, are only one-half that level, or 500,000. It appears that the charge, though appropriate at the margin, is too high in total. Consider, however, any individual firm; its marginal unit of emissions causes 100 in damages, and its first unit causes 99.9 in damages if all of the other firms continue to emit 10 units each. Thus its total damages are $10(0.5)(100 + 99.9) = 999.5$, essentially the same as its charge payments.

This result is really no different than what happens with any factor that is not in perfectly elastic supply. Consider the analogy of the labor market. According to the standard neoclassical model the wage rate is determined by the marginal worker. If the supply of labor is not perfectly elastic, inframarginal workers reap rents, and the total wage bill in the industry exceeds the sum of individual workers' reservation wages. Some firms may go out of business, although they would be profitable if they could hire some inframarginal workers at those workers' reservation wages, which are below the equilibrium wage. Yet clearly it is not inefficient for those firms to disappear. The opportunity cost of those inframarginal workers is not their reservation wages, but rather their marginal revenue product, which is equal to the market wage. Thus as long as the firm's use of labor is small relative to the relevant market, its total wage bill is an accurate reflection of the total social cost of the workers it employs.

Rose-Ackerman's argument is valid if a firm's emissions comprise a significant fraction of the total, for in that case its charges may exceed the total damages it causes by a significant amount. A simple uniform charge per unit of emissions is then inappropriate; the marginal rate should vary with emissions and marginal damages. As Pratt and Zeckhauser (1981) show, each firm's total charges should equal the difference between total damages with and without the firm in production.[3]

Revenues and Efficiency

Thus far we have ignored the efficiency implications of the effects of alternative regulatory approaches on government revenues and expenditures, except insofar as they affect the price of the product. Emission charges raise revenues, as do marketable permits if they are auctioned rather than allocated free of charge. Standards, leaving aside enforcement costs, create neither revenues nor expenditures. Subsidies may require significant expenditures. Most studies of regulation treat these effects as pure transfers, with implications for distribution but not for efficiency.

Authors who have discussed the revenues raised by charges frequently have viewed them as a problem rather than an opportunity. Anderson et al. (1977), for example, express concern about "revenue addiction"; the government may view pollution charges primarily as a method of raising revenues to be manipulated for purposes wholly unrelated to pollution control. Spence and Weitzman (1978) note that "Analysts have worried that the effluent charge scheme . . . possesses some embarrassing features on the financial side. It generates a lot of revenue that then has to be disbursed. And, effluent sources pay double; once for cleanup and once for the effluent fees on emissions after cleaning up." In a similar vein Drayton (1978) asks, "Does it make sense or is it fair to add the cost of effluent fees to the burden of cleaning up?"

The argument that charges are undesirable because "companies pay twice" has little merit. Firms do not pay twice for the same emissions. Their cleanup costs cover the pollution they eliminate, whereas the fees cover the pollution they choose not to eliminate. If we think of the environment as a scarce resource that can be put to alternative uses, the point becomes clear. Consider another scarce input, labor. Over time, as real wage rates rise, firms modify their production techniques where possible to reduce labor inputs. By analogy to the pollution case, a firm might argue that it was paying twice for higher wage rates, once for labor-saving changes in the production process and once for increased compensation for its remaining employees. A firm making such an argument might find sympathy in some quarters but it would be hard pressed to find responsible economists to support a plea for government aid to offset the effects of higher wage costs.

In the case of pollution firms may argue that they deserve financial relief because the "rules of the game" have changed, causing them to lose a valuable property right, the opportunity to emit free of charge. Note, however, that standards and permits also change the rules and impose costs on firms, and under those schemes as well a case could be made on equity grounds for some form of short-run relief for existing firms.

Some authors, including Kneese and Bower (1968), have argued that far from regretting the revenues raised by emission charges, economists interested in efficiency should welcome them to the extent that they displace existing taxes or permit the expansion of desired government programs without the need for tax increases. The public finance literature is replete with analyses of the deadweight efficiency losses associated with almost every tax in use.[4] Charges on externalities are virtually unique among taxes in reducing rather than exacerbating distortions. Thus, if revenues from pollution-reduction incentive schemes are substituted for other taxes, a net benefit is reaped. To the extent that such schemes are modified to reduce or eliminate revenues, this efficiency gain is lost. This argument suggests a further inefficiency with subsidies; they must be financed by an increase in taxes, with attendant efficiency losses.

Recognition of the efficiency implications of the level of revenue raised suggests two interrelated lines of inquiry: (1) the extent to which the optimal charge rate may diverge from marginal damage and (2) the magnitude of the tax-displacement benefit of charges relative to the efficiency gain from a cost-effective allocation of control efforts. In the analysis that follows the explicit comparison is between charges and standards. The reader should keep in mind, however, that the results for charges hold as well for marketable permit schemes that raise the same amount of revenue.

Optimal Charge Rate
The objective is to find the charge rate that minimizes social cost, defined in this context as control costs plus residual damages, less tax-displacement benefits. Using notation introduced earlier, the problem is to select t to minimize social cost,

$$S = \sum_{i=1}^{n} C^i(x_i) + \lambda_i \sum_{i=1}^{n} x_i - \delta t \sum_{i=1}^{n} x_i, \tag{3.2}$$

which is the sum of control costs and residual damages, as before, minus the tax-displacement benefit. The latter is the marginal deadweight loss per dollar of revenue collected under the displaced tax (δ) times the revenues raised by the emission charge ($t \sum_{i=1}^{n} x_i$). Note that the x_i's are implicit functions of t. As shown in chapter 2, if the tax-displacement effect is not included, the optimal charge rate is equal to marginal damage, λ. To find the optimal charge rate in the present case, we differentiate equation (3.2) with respect to t:

$$\frac{dS}{dt} = 0 = \sum_{i=1}^{n} (C_x^i)\frac{dx_i}{dt} + (\lambda - \delta t) \sum_{i=1}^{n} \frac{dx_i}{dt} - \delta \sum_{i=1}^{n} x_i. \tag{3.3}$$

Given cost-minimizing behavior by firms, $-C_x^i = t$ ($i = 1, \ldots, n$). Note that $\sum_{i=1}^{n} dx_i/dt$ is simply the inverse of the slope of the aggregate demand curve for the right to emit. Thus

$$\sum_{i=1}^{n} \frac{dx_i}{dt} = \frac{-\varepsilon \left(\sum_{i=1}^{n} x_i \right)}{t}, \tag{3.4}$$

where ε is (the absolute value of) the aggregate elasticity of demand for emissions with respect to the charge rate. Substituting $t = -C_x^i$ and equation (3.4) in equation (3.3) yields the optimal tax rate:

$$t^* = \frac{\lambda}{1 + \delta(1 - 1/\varepsilon)}. \tag{3.5}$$

Note that if there is no deadweight loss associated with the taxes displaced by the charge revenues ($\delta = 0$), the optimal charge rate is equal to marginal damage ($t^* = \lambda$), as before. Similarly, if the demand for emissions is unit elastic ($\varepsilon = 1$), the charge rate is also the same as before, because the goals of maximizing revenues and minimizing the sum of control costs and damages are not in conflict; all charge rates yield the same revenue. If $\delta > 0$ and $\varepsilon \neq 1$, however, the optimal charge rate is not equal to marginal damage. Specifically, if $\varepsilon > 1$, then $t^* < \lambda$, and if $\varepsilon < 1$, then $t^* > \lambda$. Intuitively, if demand for the right to emit is elastic, charge revenues will increase as the charge rate is lowered below marginal damage; if demand is inelastic, raising the charge rate above marginal damage will increase revenues.[5]

Terkla (1979), as part of his study *The Revenue Capacity of Effluent Charges*, develops estimates of δ for both personal and corporate income taxes. His estimates, based on work by Browning (1976), Feldstein (1978), and others, suggest that the value of δ may be quite large. For income and social security taxes on labor income, his middle estimate is $\delta = 0.35$, though that estimate is very sensitive to assumptions about the compensated supply elasticity for labor. His estimate for the deadweight loss due to taxes on capital income is still larger, $\delta = 0.60$. These estimates may strike many readers as unreasonably high, but note that they are, appropriately for our purposes, estimates of the *marginal* losses from those taxes; the corresponding estimates of average losses would be far lower, because the "welfare-triangle" loss from tax-induced distortions rises with the square of the tax rate.

Figure 3.4 plots the percentage deviation of the optimal emission charge rate from marginal damage as a function of δ for several values of ε. Note that the effects of varying ε about unity are quite asymmetric. As

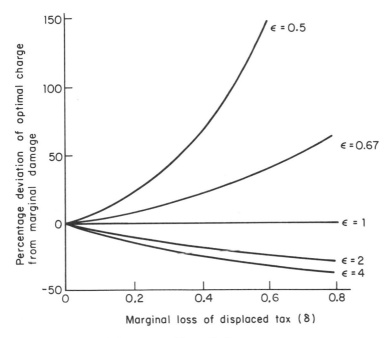

Figure 3.4   Optimal charge rate with tax displacement

demand becomes more inelastic, the optimal charge rate rises sharply with δ, but as it becomes more elastic, the optimal charge rate never falls much below marginal damage. For example, if $\delta = 0.5$, the optimal charge rate is 100 percent higher than marginal damage if $\varepsilon = 0.5$, but only 20 percent lower if $\varepsilon = 2$. The more elastic demand is, however, the larger the change in the quantity of emissions for any given percentage change in price. Thus, with a constant elasticity demand curve, a 100 percent increase in the charge rate decreases the quantity of emissions by only 29.3 percent for $\varepsilon = 0.5$, whereas a 20 percent decrease in price increases emissions by 56.3 percent for $\varepsilon = 2.$[6] Under a marketable permits scheme the regulator would adjust the number of permits to be sold so as to increase revenues.

The reader should note that in deriving the optimal charge rate we have ignored its effect on the price and quantity of the final good. To the extent that the charge deviates from marginal damage, the price of the product will not be equal to marginal social cost, resulting in the distortions discussed in the previous section. The magnitude of the efficiency losses due to such distortions depends on the proportional effect of the charge on price and on the elasticity of demand for the final good. The larger either of those variables, the less the truly optimal charge will diverge from marginal damage.

Magnitude of Tax-Displacement Gain
In estimating the magnitude of the tax-displacement benefits, we proceed in two steps, first calculating the gain if the charge rate is held at the level of marginal damage and then deriving the additional gain if the charge rate is optimized in the manner described. The results suggest that the first effect may be quite large, whereas the second is negligible.

Once again let us assume that control costs take the form $C^i(x_i) = a_i x_i^{-\alpha}$. As shown earlier, in equation (2.14), the sum of control costs and residual damages when the charge rate is set equal to marginal damage, $t = \lambda$, is given by

$$S^* = \left[\left(\frac{\lambda}{a}\right)^{\varepsilon}(1+\alpha)n\right]E(a^{\varepsilon}). \tag{3.6}$$

Damages comprise the fraction $\alpha\varepsilon$ of the total. As charge revenues equal residual damages, inclusion of the tax displacement effects (holding fixed $t = \lambda$) reduces social cost under the charge by the fraction $\alpha\varepsilon\delta$. For example, if $\delta = 0.4$ and $\alpha = 1$ (and hence $\varepsilon = 0.5$), the tax-displacement benefit is equal to 20 percent of the sum of control costs and damages. Thus, if we include the tax-displacement benefit, still holding the charge rate fixed at $t = \lambda$, social cost is given by

$$S = \left\{\left(\frac{\lambda}{a}\right)^{\varepsilon}[1+\alpha(1-\delta)]n\right\}E(a^{\varepsilon}). \tag{3.7}$$

As derived earlier, equation (3.5), the optimal charge rate taking account of tax-displacement effects is $t^* = \lambda/[1 + \delta(1 - 1/\varepsilon)]$. Recall that for the control-cost function being used, the elasticity of demand for emissions is $\varepsilon = 1/(1 + \alpha)$. Thus the optimal charge rate is

$$t^* = \frac{\lambda}{1 - \alpha\delta}. \tag{3.8}$$

Employing this charge rate, it can be shown that the net social cost, including the tax-displacement benefit, under the charge is

$$S = \left\{\left(\frac{\lambda}{a}\right)^{\alpha\varepsilon}(1+\alpha)(1-\alpha\delta)^{\varepsilon}n\right\}E(a^{\varepsilon}), \tag{3.9}$$

which is identical to the expression for social cost when the tax-

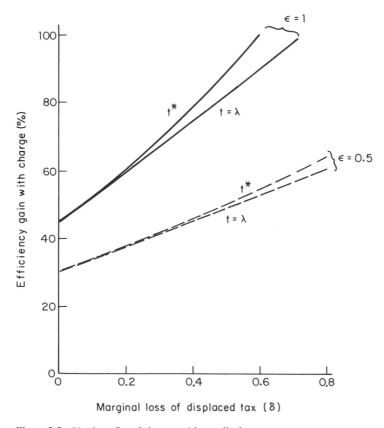

Figure 3.5   Net benefits of charges with tax displacement

displacement effect is ignored, equation (3.6), except for the multiplicative term $(1 - \alpha\delta)^\varepsilon$. For α, $\delta > 0$, $(1 - \alpha\delta)^\varepsilon < 1$. If $\alpha = 1$ and $\delta = 0.4$, for example, this term takes on the value 0.775. That is, for these parameter values optimizing the charge rate increases the net tax-displacement saving from the 20.0 percent calculated previously to 22.5 percent, a modest incremental improvement.

If as before we assume that the cost coefficient, a, is distributed log normally, we can derive expressions for the savings in social cost, including the tax-displacement benefits, of shifting from a standard to a charge. Also as before let us express these savings as a proportion of the cost of control under the standard. With the charge set equal to marginal damage, the relative gain is given by

$$\Delta S = (1 + \alpha)[1 - (1 - \alpha\varepsilon\delta)\exp(-0.5\alpha\varepsilon^2\sigma_a^2)]. \tag{3.10}$$

If the charge is optimized to take account of tax-displacement effects, the relative gain is

$$\Delta S = (1 + \alpha)[1 - (1 - \alpha\delta)^\varepsilon \exp(-0.5\alpha\varepsilon^2\sigma_a^2)]. \tag{3.11}$$

Note that equations (3.10) and (3.11) are both identical to equation (2.23) if $\delta = 0$.

Figure 3.5 plots the relative savings as a function of δ for $\sigma_a^2 = 2$ and $\alpha = 1$ or $\alpha = 0.5$. As the graph shows, the tax displacement benefits are significant for larger values of δ, whether or not the charge rate is optimized. If, for example, $\alpha = 1$ and $\delta = 0.4$, the difference in social costs amounts to more than 75 percent of the cost of control under the standard, as compared to 44.2 percent when tax-displacement effects are not included. If $\delta = 0.6$, Terkla's estimate for the marginal deadweight loss of taxes on capital, the total savings with the charge equal to marginal damage is 91.0 percent, more than double the base amount (i.e., the tax-displacement benefits of switching to the charge actually exceed the benefits from reduced damages and control costs). The results for $\alpha = 0.5$ are less dramatic but still large on a proportional basis.[7] Note, however, that for neither value of α are the additional gains from optimizing the charge rate (in both cases raising it above marginal damage) very large. Moreover, as discussed earlier, if we take account of the fact that charges that deviate from marginal damage distort the prices of final goods, even these small apparent gains would be reduced.

These results indicate that the revenue implications of alternative instruments deserve more attention than they have received in the literature and in nonacademic debates. If the revenues raised by charges are used to reduce existing distorting taxes, the benefits may be roughly on a par with the gains derived from a more efficient allocation of control efforts. Marketable permits offer the same advantage if they are sold by the government.

The results also suggest that various proposals to minimize the revenues under charge or permit schemes are misguided. Some authors, for example, have suggested that a benchmark be established, with charges levied on emissions above that level and subsidies paid for reductions below it, with the benchmark and the charge/subsidy rate adjusted to balance charge receipts and subsidy payments. Similarly, some authors have proposed that permits initially be allocated free of charge rather than sold.[8] Such approaches eliminate the tax-displacement benefit discussed here. At the same time, however, our results suggest that little is gained (or lost) by having regulators try to adjust charge rates (or the number of permits) away from marginal damage in order to increase revenues.

The revenues raised by a charge system or through the sale of marketable permits could be used for a variety of purposes. On both political and economic grounds two options appear especially promising. The first would be to use the revenues to provide aid to the communities where the risks are suffered. Such a scheme would provide a form of compensation and thus could be promoted on equity grounds. It also would provide significant efficiency gains, as the taxes displaced at local levels are likely to be particularly inefficient.

A second option would be to use the revenues to reduce corporate income tax rates. That plan also might find political support on equity grounds, as the charge revenues would be raised primarily from the corporate sector.[9] Given the relatively high deadweight losses associated with taxes on capital income, that approach also can be recommended on the basis of efficiency.

Summary

In this chapter and the previous one the performances of several alternative incentive mechanisms have been compared with those of standards along three different dimensions: (1) the degrees to which they allocate control efforts efficiently across sources, (2) their impacts on the prices of final goods, and (3) the extent to which they provide additional benefits by displacing distorting taxes with their accompanying deadweight losses. Table 3.2 summarizes the findings, with a " + " indicating performance superior to the standard, a " − " indicating inferior performance, and a "0" indicating no change in performance.

As shown in chapter 2 and as indicated in the first column of table 3.2, any of the incentive mechanisms considered outperform standards in terms of allocating control efforts efficiently; all allocate control efforts so as to equalize marginal control costs across sources. Charges and marketable permits (either auctioned or allocated free of charge) also exert the appropriate impact on final-goods prices, as indicated in column 2, incorporating residual external damages in the price of the product. In contrast, a subsidy lowers the price of the product below the level achieved under a standard, thus encouraging excessive consumption and raising overall emissions.

As shown in the last column, the subsidy also suffers from the fault

Table 3.2 Qualitative summary of findings

	Efficient control	Appropriate product price	Tax-displacement benefit
Direct regulation			
Standard	0	0	0
Incentive mechanisms			
Charge	+	+	+
Marketable permit			
Auctioned	+	+	+
Allocated	+	+	0
Subsidy	+	−	−

Note: 0 indicates instrument performs as well as standard; " − " indicates instrument performs worse than standard; " + " indicates instrument performs better than standard.

that it requires additional government expenditures, thus necessitating increases in distorting taxes or cutbacks in existing programs. In contrast, the charge approach generates new revenues, allowing tax reductions or program expansion. Marketable permits yield the same benefits, but only if they are sold; if they are allocated free of charge, these benefits are not reaped.

Our results thus far do not allow us to distinguish between charges and marketable permits that are sold. Both approaches allocate control efforts in a least-cost manner, both cause residual damages to be reflected in final product prices, and both generate tax-displacement benefits. None of the other instruments considered—standards, subsidies, or marketable permits allocated free of cost—does as well on all dimensions. As shown in the next chapter, however, charges and permits are not equivalent once we consider the effects of uncertainty and change over time.

Chapter 4

Dynamic Efficiency: Uncertainty and Change

Uncertainty and change pervade environmental regulation. Any analysis that ignores these issues is at best incomplete, and at worst may be seriously misleading. In the two preceding chapters we showed that emission charges and marketable permits initially allocated by auction yield identical results and are superior to uniform standards in a static world where the benefits and costs are known with certainty. This chapter analyzes the impacts of uncertainty and change on the relative efficiencies of alternative instruments. Given the results of the previous chapter, subsidies will not be considered.

The most striking uncertainties arise in trying to estimate the benefits of reduced emissions. Our understanding of the links between emissions and damages is primitive; for most pollutants, available risk "estimates" could more accurately be characterized as "educated guesses." Moreover techniques for assigning monetary values to reductions in risk are neither well developed nor widely accepted. Estimates of control costs also are highly uncertain; reliable data are difficult and expensive to obtain. Typically, regulatory agencies must rely on information from the affected industry, which may have a strong incentive to misrepresent costs in an effort to avoid regulation, and on limited engineering analyses of model plants by consultants.

These pervasive uncertainties raise two major issues for the evaluation of alternative regulatory instruments. First, what should be the decision criteria? Should expected costs and benefits be employed, or is some other, more conservative approach, such as worst-case analysis, more appropriate? Second, how robust are the different regulatory instruments in the face of significant uncertainties? A policy whose performance is highly sensitive to the accuracy of prospective cost and benefit estimates is unlikely to perform well in practice. Ideally, the policy should incorporate mechanisms for self-correction as uncertainties are resolved.

The problems of change compound the problems of uncertainty. The regulator must deal with a "moving target," and the performance of a policy that works well initially may deteriorate significantly over time as costs and benefits change. The benefits of control, for example, may rise as population densities increase, or costs may fall as firms develop innovative control technologies. These factors reinforce the importance of designing policies that are self-correcting. At the very least the regulatory strategy should be able to accommodate beneficial change. Better yet, it should encourage and accelerate it.

Decision Criteria under Uncertainty

Regulators must operate in a highly uncertain environment. Prospective cost estimates, for the reasons cited earlier, are likely to be quite rough. The benefit estimates tend to be cruder yet, because of uncertainties about both the physical consequences and the appropriate way to value those consequences. Plausible benefit estimates for any particular regulation may cover several orders of magnitude, with some estimates suggesting little or no risk, whereas others indicate potentially disastrous results if stringent controls are not imposed.

Faced with these uncertainties, the decision maker has several options, including (1) using expected costs and benefits as the bases for decision making and (2) taking a conservative approach that emphasizes worst-case outcomes. Each approach has many variants, depending on its precise definition. The second is particularly sensitive to definitions; worst case can cover a wide range of possibilities. The definition of expected costs and benefits is conceptually sharper, though often difficult to apply given the fuzzy probabilities involved.

The Expected Value Criterion
Under the expected value criterion the regulator seeks to maximize expected net benefits: expected benefits minus expected costs. That is, the regulator assesses the range of possible outcomes under each alternative and then computes the weighted average of costs and benefits, where the weights are the probabilities associated with each outcome.

The expected net benefit criterion has been criticized sometimes for ignoring large asymmetries in costs and benefits, in particular the fact that the potential losses from not regulating may be much greater than the costs of regulating. Consider the following simple example. EPA is considering banning a particular industrial chemical for which more expensive (but safe) substitutes are available. The probability that the chemical is harmless is 0.9, in which case the relatively modest costs of banning the chemical will, in essence, have been wasted. There is, however, a 0.1 probability that the chemical is a potent carcinogen, in which case the benefits of the ban would swamp the costs.

A decision criterion that relied on the "best" or "most likely" risk estimate would reject the ban, because there is only a small chance that it would yield positive net benefits. The asymmetry in the loss from making a "wrong" decision would be ignored. Note, however, that the expected net benefit criterion does not ignore this asymmetry, as it incorporates both the magnitudes of the costs and benefits and their probabilities. In this simple example the regulator would ban the chemical if the potential benefits were at least ten times the cost of the ban. Thus the expected value criterion should not be confused with approaches that ignore uncertainties and use only the "most probable" parameter values.[1]

Zeckhauser (1979) discusses a closely related problem. He notes that, in estimating low-level risks, scientists may report the median or mode as the "best" estimate. If the subjective probability distribution is normal (or follows some other symmetrical distribution), that estimate also gives the expected level of risk. If, however, the distribution is not symmetrical, the median risk estimate may be quite different than the mean. Suppose, for example, that a scientist reports that his "most likely" estimate of risk is 10^{-4} but that it could be 10^{-3} or 10^{-5}. More specifically, he thinks that there is a 0.5 chance that the risk is 10^{-4}, with the higher and lower estimates each having a probability of 0.25. The expected risk then is not 10^{-4} but rather $0.25(10^{-3}) + 0.5(10^{-4}) + 0.25(10^{-5}) = 3.025 \times 10^{-4}$, which is higher by a factor of three.

Even if the expected values of individual parameters are used to estimate costs and benefits, the resulting estimates may not represent expected values if the costs and benefits are nonlinear functions of the individual parameters. Suppose, for example, that risk is proportional to the dose squared. That is, $R = \gamma D^2$, where R is the risk, D is the dose in appropriate units, and γ is a constant. If we look across the potentially affected population, the dose is equally likely to be 0, 5, or 10. Thus the expected dose is 5, and one might be tempted to conclude that the expected risk is $\hat{R} = \gamma(5^2) = 25\gamma$. In fact, however, the true expected risk is substantially higher, $\hat{R} = \gamma[(10^2 + 5^2 + 10^2)/3] = 40.67\gamma$.

Risk Aversion
A more sophisticated criticism of the expected net benefit criterion is based on the concept of risk aversion. It is well established that most individuals, if given a choice, will not accept actuarially fair gambles involving substantial sums of money (Raiffa 1968). Few people, for example, would be willing to enter a lottery that gave them a 50-50 chance of doubling their annual income or of losing it altogether, although under the expected value criterion they should be indifferent, as the expected value of the lottery is zero.

The analogy to decisions about risks to health is obvious. Often, however, the concept of risk aversion is misinterpreted or misapplied. Consider the example begun earlier. Suppose that if the chemical, call it A, is a carcinogen, individuals exposed to it run a 10^{-3} incremental risk of developing cancer. If it is safe, then the incremental risk is 0. Given the 10 percent chance that chemical A is carcinogenic, the expected risk from exposure is $0.9(0) + 0.1(10^{-3}) = 10^{-4}$. Chemical B is a proven, but weaker, carcinogen, known to pose a risk of 10^{-4}. Decision analysis tells us that an individual unable to gather additional information should be indifferent between the risks posed by the two chemicals. Raiffa 1968, 120) refers to this issue as "the reduction of a general lottery." Risk aversion does not enter into the decision at all; from an ex ante perspective both chemicals pose exactly the same risk. (From the ultimate ex post perspective, for any individual both will also pose the same risk, either zero or one.)

From a societal perspective, however, some would argue we should treat the two chemicals differently. Suppose one million (10^6) people are to be exposed. By the law of large numbers we can be virtually certain that with chemical B close to $10^{-4}(10^6) = 100$ people will contract cancer. With chemical A there is a 0.9 probability that no one will contract cancer, but a 0.1 probability that about $10^{-3}(10^6) = 1,000$ will. If we are risk averse about the number of lives lost, we should treat A as a more serious problem than B; in some sense it may be more than ten times as bad to have 1,000 cases of cancer associated with some chemical than to have 100.

This situation should be distinguished from a similar one that is often raised in connection with nuclear power and other technologies that pose the possibility of large-scale catastrophies killing large, geographically concentrated groups of individuals. In that connection some observers argue that it is worse for 1,000 people to die in one incident than for the same number to die in separate incidents scattered around the country.[2] Even in that case, however, it is far from clear that marginal damage increases with the number of fatalities; cogent arguments have been made suggesting that marginal damages actually may decline.[3] Whatever the merits of these competing arguments, the situation with regard to environmental carcinogens is quite different; even with highly potent carcinogens the impact is likely to be spread over time and space, and there seems no reason to assign different values per case of cancer depending on the number of cases that can be attributed to a particular chemical. The risk posed by any single chemical almost certainly will be small relative either to the risk caused by all chemicals or to the total risk of death faced by individuals; hence, even if our valuation function is nonlinear for total deaths, it will be essentially linear over the limited range relevant to any single decision.[4]

The fact that we can assume, in general, that the dollar benefits will be effectively a linear function of the adverse health effects prevented simplifies the analysis considerably. First, it means that the expected values of the levels of the effects will be sufficient for making most decisions. Second, it means that the excruciatingly sensitive problem of assigning dollar values to health effects can be carried out at the last stage of the analysis. Alternative regulatory policies can be summarized in terms of expected costs and expected health gains. In many cases all but a few potential policies may be dominated by others, thus narrowing the field of choice considerably. The remaining policies may then be compared in terms of cost-effectiveness ratios, or explicit dollar values may be assigned to "saving a life" or preventing a case of cancer. At that point sensitivity analyses of alternative weighting schemes may be performed easily.

Conservative or Worst-Case Analysis

The major competitor to the expected value criterion in terms of frequency of use is conservative or worst-case analysis. Under that approach the analyst selects those assumptions and parameter values that give high estimates of risk. One obvious conceptual difficulty with this approach is that the worst case may be hard to define. There is, for example, some finite, though infinitesimal, probability that all leukemia deaths are the result of benzene exposure, that if we eliminated all benzene from the environment, leukemia would disappear.[5]

Usually a less extreme, "plausible," worst case is constructed. In some instances the criterion employed may be quite explicit. In a classic article on safety testing of carcinogens, for example, Mantel and Bryan (1961) propose that, when estimating risk from animal studies, the upper 95 percent confidence estimate should be used rather than the observed or expected value. More typically, however, the criteria are much less explicit and, perhaps, less obvious. For example, dispersion modeling conducted for EPA's proposed standard for benzene emissions from maleic anhydride plants employed meteorological data from Pittsburgh, where the draft environmental impact statement notes, "meteorological conditions that maximize ground level concentrations ... are common" (U.S. EPA 1980, 4–11). Similarly, as discussed in chapter 8, EPA's estimates of benzene emissions from chemical plants were based on the assumption that all plants operate at full capacity all the time, an assumption that clearly biases upward the estimates of emissions and hence of risk.

Conservatism also may be introduced through the choice of models rather than through the selection of specific parameters. In extrapolating from high-dose epidemiological data on benzene exposure to the low doses found in the ambient environment, for example, EPA has relied on the linear dose-response model, under which risk is proportional to dose. As discussed more fully in chapter 8, that model predicts much higher risks at low doses than do alternative models. As the preliminary report of EPA's Carcinogen Assessment Group stated, "such a model is expected to give an upper limit to the estimated risk" (Albert et al. 1977, 1).[6]

Several rationales can be offered for performing conservative or worst-case analyses. None, however, is very satisfactory. As Zeckhauser (1975) notes, academics often make conservative assumptions unfavorable to their own positions in order to bolster their arguments; an opponent of some project might attempt to show that, even if the benefits of the project were much higher than predicted and its costs much lower, the project still would fail to yield positive net benefits. Thus, for example, it might make sense for EPA to do a preliminary worst-case analysis of a particular chemical to see if further investigation were warranted. Note, however, that such an analysis logically can prove only the case for not regulating. It cannot, in general, provide a strong case for regulation because the assumptions purposely are chosen to bias the risk estimate upward. When agencies use "conservative" risk analyses in support of their regulations, they use them in just the opposite way such analyses usually are employed in academic discourse.[7]

Conservative analyses also may be defended as providing a margin of safety; it is better to err on the side of safety. "Better to be safe than sorry." Conservative analysis may be thought of as a way of giving greater weight to risk reduction relative to control costs. Biasing risk estimates upward, however, is an arbitrary and inappropriate way to accomplish that goal. In a world of scarce resources conservative approaches distort the trade-offs that decision makers face and may lead to less rather than more safety, particularly when they must compare the return to control expenditures in different areas where the degrees of uncertainty vary.

Consider the following simple example. Suppose that EPA must decide whether or not to ban chemical C. If it does, chemical D is the only available substitute and will be used. The risk with chemical D is known to be 10^{-5} per person exposed. The risk with chemical C is less certain.

Scientists estimate that there is a 95 percent chance that it is harmless but a 5 percent chance that it poses a risk of 10^{-4}. If EPA employed worst-case analysis, chemical C would be banned, as $10^{-4} > 10^{-5}$. The ban, however, would lead to higher risk, on average, as the expected risk with C is only $0.95(0) + 0.05(10^{-4}) = 0.5 \times 10^{-5}$, half the risk associated with D.[8]

The major problem with conservative or worst-case analysis is that it distorts the information provided decision makers in unpredictable and often obscure ways. As noted earlier, the criteria for deriving a worst-case scenario are rarely made explicit, and implicit criteria are likely to vary widely depending on the problem and the analyst. Thus decision makers cannot really be sure what they are getting. Does the estimate represent the analyst's 90, 95, or 99.9 percent upper confidence limit? Is it 2, 5, 10, or 100 times higher than the expected value?

If the estimate is the product of several intermediate stages, at each of which conservative assumptions have been injected, the final estimate may be far more conservative than intended. To predict the risk from a particular chemical, for example, EPA must estimate three different factors: (1) the level of emissions, (2) the level of exposure per unit of emissions, based on dispersion in the atmosphere and population patterns, and (3) the level of risk per unit of exposure (the dose-response function). The final risk estimate is the product of these three factors.

Suppose that at each stage EPA's analyst uses the 90 percent upper confidence limit, which, let us further assume, is two times the expected value of that parameter. If the distributions of the individual parameters are independent, however, the final risk estimate will not be double the expected value but rather $2^3 = 8$ times higher.[9] Thus a series of plausible worst-case assumptions can compound to yield a highly implausible estimate of risk. Moreover the degree of conservatism in the final estimate is dependent on the number of stages into which the problem has been decomposed, and its magnitude may well go unrecognized, as each conservative assumption taken in isolation may appear quite modest.

Summary
Worst-case or conservative analyses of risk do not provide an appropriate basis for deciding to regulate. Such analyses distort the trade-offs involved and may lead the decision maker seriously astray. They may even lead to higher rather than lower risk. Thus, other than to prove the negative (i.e., that regulation is not desirable), conservative approaches to risk analysis should be avoided. In general, expected costs and benefits (reduc-

tions in risk) provide the best bases for decision making. When applied correctly, the expected value approach gives appropriate weight to both the probabilities and the consequences of alternative outcomes.

The expected value criterion, despite its firm conceptual basis, may be difficult to implement. Most of the uncertainties associated with cost and risk estimates defy ready quantification. Faced with an array of dose-response models that predict very different risk levels, for example, the appropriate conceptual approach is to take a weighted average of the estimates, where the weight attached to each estimate reflects the probability that the associated model is correct. Unfortunately, we have no well-defined way to assign such probabilities. One possibility might be to get a distinguished panel of scientists, of the type frequently assembled by the National Academy of Sciences, to develop these probabilities using the Delphi technique or a similar procedure.[10] Such an exercise would be difficult to conduct and probably would not yield very firm, widely accepted estimates. It would, however, be a major step forward.

Despite the uncertainties involved, many sources of conservatism in current analyses can be eliminated. Actual rather than full-capacity production levels, for example, should be used to estimate emissions. Similarly, the inputs to dispersion models should be typical of meteorological conditions at the sites in question, rather than chosen to maximize predicted concentrations. Alternatively, sensitivity analyses employing a range of assumptions can be performed, giving the decision maker an idea of the range of plausible estimates rather than simply one extreme.

Robustness under Uncertainty

Regulations should be designed to maximize expected net benefits. Because of major uncertainties, however, the actual costs and benefits may be far different than those predicted, and more important, the regulation chosen using imperfect information may be very inefficient. Improving the information on which regulatory decisions are made is one way to increase performance, but better information is often costly, and even the best data will leave major uncertainties. Thus regulators do well to choose regulatory strategies that are responsive to uncertainty, providing some built-in mechanisms for self-correction.

Perhaps the most telling argument against emission charges has been that, at least in the classical formulation, the regulator must know the level of marginal damage at the optimum. But typically we have only the crudest estimates of damages. Moreover, if marginal damage varies with the level of emissions, the regulator also needs an accurate estimate of the marginal cost curve. Under such real-world conditions, critics conclude, charges are totally impractical.

Target Emission Levels

In the face of this criticism some advocates of charges have retreated from the classical prescription of setting the charge equal to marginal damage. Baumol and Oates (1971), for example, advocate a hybrid approach, whereby the government would set a target level of pollution (an ambient standard), which it would then achieve through emission charges. They are disturbingly vague as to how such a target would be set, but the same problem confronts regulators setting an ambient target to be achieved through emission standards. Under the Baumol-Oates plan the regulator would then set the charge based on estimated marginal control costs at the target level. The result would not necessarily be optimal, but it would yield the target at minimum cost.

If the cost estimates are inaccurate, however, the ambient target will not be achieved. In particular, if costs are higher than estimated, environmental quality will be lower than desired. Conversely, if costs are lower than predicted, firms will spend more on control than is required to meet the overall goal. One possible strategy is to set a tentative charge and then adjust it up or down in response to the behavior of firms (Kneese and Bower 1968). That process might take several iterations, however, before the ambient standard was met with sufficient accuracy. Because of the capital-intensive nature of most pollution-control techniques the adjustment process could be very costly (Rose-Ackerman 1974). Moreover firms might postpone the installation of controls until the rate had stabilized, thus delaying environmental improvement and making it difficult for the regulator to know what charge to set.

Emission standards offer greater assurance that the ambient target will be met, but they do not offer perfect accuracy. Such standards are almost always stated in terms of emissions per unit of output or per unit of some input to the production process. Thus, if the level of production is different than predicted, the total level of emissions also will be different than expected. In addition, of course, costs will be higher than necessary, whatever the level of control ultimately achieved.

Marketable permits appear to offer the perfect strategy for achieving an ambient target. They provide even greater assurance than standards that the desired overall level of emissions will be reached and achieve the same efficiency as charges in allocating control efforts. Permits will suffer

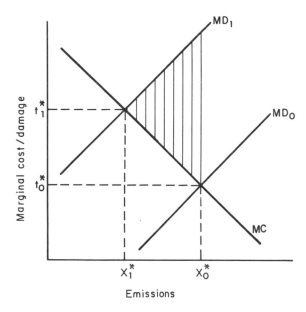

Figure 4.1   Impact of misestimating marginal damage

from some of the same problems as iterated charges, however, if the price of the permits does not stabilize quickly. More important, if actual costs or damages differ from the estimates, the original ambient target is unlikely to remain appropriate, and the ability of permits to achieve it, no matter what, may not be a virtue.

Formal Analysis

The implications of uncertainty for the choice of instruments may be explored using some simple diagrams.[11] Consider first uncertainty about the damage function, which, as discussed earlier, is likely to derive from several sources and to be substantial. Moreover unlike uncertainty about costs, it is unlikely to narrow significantly over time.

Expected social cost is minimized at the point where expected marginal damage is equal to marginal cost. Figure 4.1 illustrates this condition, where aggregate marginal cost is represented by the curve labeled MC and expected marginal damage is MD_0. Ex ante, t_0^* is the optimal charge rate, and X_0^* is the optimal number of permits. Suppose that actual marginal damage is greater than expected, as shown by MD_1 in figure 4.1, so that ex post the optimal level of emissions is X_1^*. Neither the charge nor the permit scheme yield that level; the number of permits fixes emis-

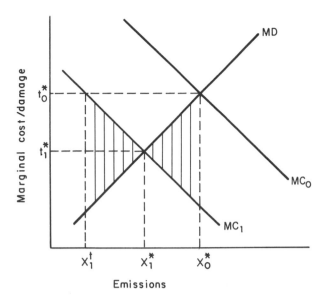

Figure 4.2   Impact of misestimating marginal costs

sions at X_0^*, and the charge also yields that same result because the level of emissions is determined by the intersection of the charge and the marginal cost curve. Both approaches lead to excessive emissions, and social cost exceeds its minimum possible level by the area of the shaded triangle. If marginal costs were lower than expected, both approaches would lead to too much control (too few emissions); the welfare loss would again be the same for both approaches. Note that, although our graphical analysis does not include standards, the same qualitative argument applies; misestimates of marginal damage will lead the regulator to set the wrong standard. Thus highly imperfect information about the damage function makes it difficult to set the right charge, but it poses equally severe problems in setting the right quantity, and hence does not provide an argument for using permits or standards instead of charges.

The relative efficiencies of charges and permits, however, may be affected by uncertainty about control costs. The uncertainties associated with control cost estimates are likely to be less severe than those associated with damages, yet still large. In figure 4.2, MC_0 is the expected marginal cost function estimated by the regulator, whereas MD is the (expected) marginal damage function. Ex ante, the optimal charge is t_0^* and the optimal number of permits is X_0^*. If the true marginal cost curve is lower, however, as shown by MC_1 in the figure, the optimal emission level is lower (X_1^*), as is the optimal charge rate (t_1^*). Under the permit approach, emissions are higher than optimal, resulting in the welfare loss (relative to the optimum) shown by the shaded triangle to the right of X_1^*.

Under a charge approach the situation is a bit more complicated. If firms make the same cost estimates as the regulator, and do not realize their mistake until after they have locked themselves into particular control levels, the result may be the same as with the permits. If, however, as seems more likely, each firm estimates its own control costs reasonably accurately, the charge t_0^* will result in emissions at the level X_1^t, which is lower than optimal given the true aggregate marginal cost function. The resulting welfare loss, relative to optimal control, is shown by the shaded triangle to the left of X_1^*. If the true marginal cost curve is higher than expected, the results are simply reversed; the marketable permits approach results in too much control, and the charge in too many emissions.

This differential impact of uncertainty about control costs has some interesting implications for the behavior of interest groups participating in the regulatory process. Under the current system of standards firms have an incentive to overestimate the costs of complying with proposed regulations in the hope that the standards will be relaxed. Conversely, environmentalists and others desiring stricter standards have an incentive to predict lower costs. The observed behavior of these groups is consistent with the incentives hypothesized; cost estimates from the affected industries almost always exceed EPA's estimates, and environmental groups typically argue that actual costs will be lower because EPA has failed to account for technological innovation (possibly accelerated by a tighter standard) or some other cost-lowering factor.[12] A marketable permits scheme preserves those incentives. Higher estimated costs lead to issuing more permits, whereas lower cost estimates lead to fewer permits. Under a charge scheme, however, the incentives are reversed. The lower the estimated marginal cost curve, the lower the charge. Hence under a charge-based approach industry representatives have an incentive to underestimate costs, whereas environmentalists have an incentive to show that control costs are high.

A more important issue is the relative magnitudes of the losses under each scheme due to misestimation of costs. The marginal cost and damage curves in figure 4.2 have been drawn so that the welfare losses are the same under both the charge and the permit approaches. It is apparent, however, that the relative sizes of the losses could differ considerably, depending on the shapes of the cost and damage functions. The steeper

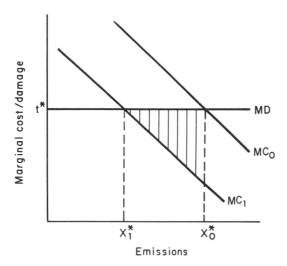

Figure 4.3   Optimal charge when marginal damage constant

the marginal cost of control curve (i.e., the more sharply nonlinear the cost of control), the more desirable is the charge approach relative to marketable permits. Conversely, the steeper the marginal damage curve (i.e., the more sharply nonlinear the damage function), the greater the relative desirability of the permit approach.

The intuitive explanation behind these results is straightforward. Consider first the shape of the cost function. If marginal cost varies little over a wide range of emission levels, then a slight error in setting the charge rate can lead to a large error in the level of emissions achieved, thus making marketable permits, which fix the quantity, more desirable. In contrast, if marginal costs are rising rapidly, small errors in setting the number of permits can lead to radical changes in control costs, whereas errors in the charge rate will have little impact on the level of emissions actually achieved.

Similar reasoning applies on the damage side. At one extreme damage is strictly linear with respect to emissions (the marginal damage curve is a horizontal line); the optimal number of permits is highly sensitive to costs, but the optimal charge rate is entirely independent of costs. Figure 4.3 illustrates the argument. As in figure 4.2, MD represents the marginal damage function while MC_0 represents the estimated marginal cost function. In this case, however, MD is constant. Ex ante, the optimal charge rate is t^*, and the optimal number of permits is X_0^*. As before, the true marginal cost curve is lower than estimated, MC_1. The optimal level of emissions falls to X_1^*. With permits emissions are fixed at X_0^*. Relative to the optimum the loss in net benefits is shown by the shaded area. Under a charge, however, the level of emissions automatically adjusts to the true optimum, and no loss is sustained. The optimal charge rate remains the same despite the change in marginal cost.

Now consider the other extreme where an absolute threshold exists; marginal damage is infinite at that point and zero elsewhere. In that case the optimal charge rate will be highly sensitive to costs, but the optimal number of permits will be independent of costs (assuming that non-marginal criteria are met; i.e., that some degree of control is desirable). Under such circumstances it is critical for the regulator to achieve the right level of emissions, and he is well advised to use permits rather than a charge.[13]

The relative curvatures of the control-cost and damage functions will vary with the type of substance involved and the control options available. In most cases, however, the aggregate cost-of-control function is likely to be sharply nonlinear. As emissions are reduced at any particular source, marginal costs often rise sharply. Even if marginal control costs are fairly constant for each source, the aggregate marginal cost function will rise rapidly as controls are tightened if costs vary widely across sources, because additional reductions in emissions will require controls at increasingly expensive sources.

The shape of the damage function is less clear and more likely to vary widely. In some cases distinct thresholds may exist or, more commonly, there may be regions over which marginal damages rise rapidly. Minor changes in the concentrations of certain water pollutants, for example, may mean the difference between a thriving fishery and a barren one (Roberts 1975). Much existing legislation appears to be based on the concept of thresholds, directing regulators to set standards at the "safe" level. Rarely, however, do real damage functions exhibit sharp jumps; more typically, disaster does not strike if the supposed threshold is exceeded, and benefits continue to be reaped as concentrations are lowered below that level.

For many environmental problems the true marginal damage function appears to be relatively constant over the relevant range of emissions. As noted earlier, for example, the most widely used dose-response model for carcinogens assumes that risk is proportional to exposure at low doses. If that model is correct, then the marginal damage curve is flat. As argued in more detail in chapter 8, even if one believes that there is only a small chance that the linear model is correct, the other (nonlinear) models predict such infinitesimal risks at low exposures that the expected dose-response function is essentially linear in the low-dose range.

Even if the true dose-response function has a threshold, expected marginal damage, the appropriate basis for decision, is likely to be continuous and fairly flat over a wide region because of uncertainties about the threshold dose. Consider a simple hypothetical example. We are certain that some threshold level of emissions, X^T, exists. If emissions exceed that level, damages are 100. If emissions are less, no damages are incurred. Marginal damage is infinite at X^T, but zero elsewhere. Thus control yields no benefits unless emissions are driven below X^T, but further reductions have no value. This appears to be the ideal situation for using marketable permits. But now suppose that we are uncertain about the value of X^T. More specifically, our subjective distribution is uniform over the interval $0 \leq X^T \leq 10$. Expected marginal damage at any level of emissions is the probability that it is the threshold times the damage that will result if the threshold is exceeded. Thus in this case the expected marginal damage function is constant at 10 over the interval $0 \leq X \leq 10$, and zero elsewhere. Under the expected value criterion we should act as if the damage function were linear over that interval, and a charge probably would be more efficient than marketable permits.[14]

This analysis suggests that uncertainty about the costs of control will increase the efficiency of charges relative to permits (or standards) in most cases. Uncertainty about the damages also increases the relative efficiency of charges to the extent that it flattens the expected marginal damage curve. Linearity of expected damages means that the regulator does not have to have any information on costs to set the optimal charge. In contrast, the regulator must have reasonably accurate cost estimates to issue even approximately the right number of permits or to set roughly the right standard. As experience with standards has shown, cost estimates can be expensive and time-consuming to develop yet still not be very reliable.[15]

A Numerical Example

If we assume that expected marginal damages are constant, we can use a modified version of the model developed in chapter 2 to derive measures of the effects of uncertainty on the relative performances of charges, permits, and standards. Total social cost under that model was given by equation (2.12):

$$S = \sum_{i=1}^{n} a_i x_i^{-\alpha} + \lambda \sum_{i=1}^{n} x_i, \qquad (4.1)$$

where the first summation represents costs and the second represents damages. As shown in chapter 2, minimum social cost (equation 2.14) can be achieved by a uniform charge on emissions at the rate $t = \lambda$:

$$S^* = \left[\left(\frac{\lambda}{\alpha} \right)^{\alpha\varepsilon} (1 + \alpha)n \right] E(a^\varepsilon). \tag{4.2}$$

Under the optimal uniform standard social cost was given by equation (2.18):

$$S_s^* = \left[\left(\frac{\lambda}{\alpha} \right)^{\alpha\varepsilon} (1 + \alpha)n \right] [E(a)]^\varepsilon. \tag{4.3}$$

The gain from shifting from the standard to the charge (or permits) was represented by the difference in social costs divided by the control costs under the optimal standard:

$$\Delta S^* = \frac{S_s^* - S^*}{\varepsilon S_s^*}. \tag{4.4}$$

These results assume that the regulator can measure marginal damage (λ) and control-cost coefficients (a_i) with perfect accuracy. Suppose, however, that λ is misestimated by a factor of k_λ. Under the charge, $t = k_\lambda \lambda$, leading the ith firm (from equation 2.4) to set its emission level at

$$x_i^+ = \left(\frac{\alpha a_i}{k_\lambda \lambda} \right)^\varepsilon = k_\lambda^{-\varepsilon} x_i^*, \tag{4.5}$$

where the "$+$" superscript indicates that the emission level is based on incorrect information. Note that if $k_\lambda > 1$, the charge is too high, and sources emit too little. Conversely, if $k_\lambda < 1$, the charge is too low, and sources emit too much. The uniform standard also will be set at the wrong level if λ is misestimated (cf., equation 2.17):

$$\bar{x}^+ = \left[\frac{\alpha E(a)}{(k_\lambda \lambda)} \right]^\varepsilon = k_\lambda^{-\varepsilon} \bar{x}^*. \tag{4.6}$$

In both cases the emission levels deviate from their optima by a factor of $k_\lambda^{-\varepsilon}$, and the total social cost for each alternative deviates from its optimal level by the following factor:

$$\frac{S^+}{S_s^*} = \frac{S^+}{S^*} = (k_\lambda^{\alpha\varepsilon} + \alpha k_\lambda^{-\varepsilon})\varepsilon. \tag{4.7}$$

For example, if $\alpha = 1$ ($\varepsilon = 0.5$) and $k_\lambda = 1/2$ (i.e., true marginal damage is twice that estimated), then social costs under each alternative increase

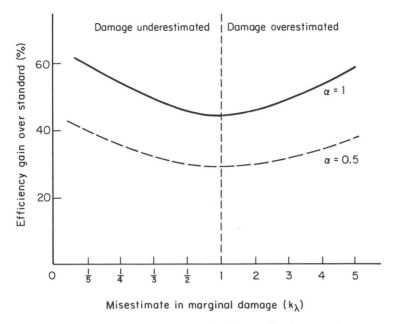

Figure 4.4   Effects of misestimating marginal damage for charges and permits vs. standard

by 6.1 percent. Relative social costs remain the same, but the absolute difference increases:

$$S_s^+ - S^+ = (k_\lambda^{\alpha\varepsilon} + \alpha k_\lambda^{-\varepsilon})\varepsilon(S_s^* - S^*). \tag{4.8}$$

Again normalizing by dividing by the control costs under the optimal standard (based on perfect information), the relative efficiency gain under the charge is

$$\Delta S^+ = \frac{(k_\lambda^{\alpha\varepsilon} + \alpha k_\lambda^{-\varepsilon})\varepsilon(S_s^* - S^*)}{\varepsilon S_s^*}$$

$$= (k_\lambda^{\alpha\varepsilon} + \alpha k_\lambda^{-\varepsilon})\varepsilon(\Delta S^*). \tag{4.9}$$

If we assume, as in chapter 2, that $\ln a$ is distributed normally, with variance σ_a^2, we can calculate ΔS^+ using the expression for ΔS^* in equation (2.23). Figure 4.4 plots the results as a function of k_λ for $\sigma_a^2 = 2$ and for $\alpha = 1$ (the solid line) and $\alpha = 5$ (the dotted line). If λ is not misestimated ($k_\lambda = 1$) and $\alpha = 1$, the charge yields a savings of 44 percent, as before. As k_λ deviates from unity, however, the savings under the charge rise, to 47 percent for $\alpha = 1$ and $k_\lambda = 2$ or $k_\lambda = 1/2$, and to 59 percent for $\alpha = 1$ and $k_\lambda = 5$ or $k_\lambda = 1/5$; for $\alpha = 1$, misestimating marginal damage by a given factor has the same effect whether the estimate is too high or too low.

For $\alpha = 0.5$, the savings are lower at all levels, and the effects of uncertainty are not quite symmetric; underestimating marginal damage by a given factor improves the relative efficiency of the charge slightly more than overestimating it by the same factor. Thus, although mistakes in estimating marginal damages lower the efficiency of both the charge and the standard, they increase the efficiency gap between the two. Note that the results for the charge also hold for marketable permits; as shown earlier in this chapter, charges and marketable permits remain equivalent under conditions of uncertainty about the damage function.

We can use the same approach to analyze the effects of misestimating control costs. Suppose that costs are misestimated by a factor of k_a; for example, if $k_a = 2$, the regulator's cost estimates for all sources are double their true values. This has no effect on the efficiency of the charge because the optimal charge rate ($t^* = \lambda$) does not depend on costs given our linear damage function. Thus $S^+ = S^*$. It does, however, reduce the efficiencies of both marketable permits and standards. If costs are misestimated by a factor of k_a, the standard will be set incorrectly (cf., equation 2.17):

$$\bar{x}^+ = \left[\frac{\alpha E(k_a a)}{\lambda}\right]^\varepsilon = k_a^\varepsilon \bar{x}^*. \tag{4.11}$$

The optimal number of permits, X^*, is equal to the sum of emissions achieved under the optimal charge. Using equation (2.13), we obtain

$$X^* = \sum_{i=1}^{n}\left[\frac{\alpha a_i}{\lambda}\right]^\varepsilon. \tag{4.12}$$

If the a_i's are misestimated by a factor of k_a, however, the number of permits issued will be

$$X^+ = \sum_{i=1}^{n}\left[\frac{\alpha k_a a_i}{\lambda}\right]^\varepsilon = k_a^\varepsilon X^*. \tag{4.13}$$

Thus, under both the standard and permits, the level of emissions will deviate from the optimum by a factor of k_a^ε.

The social costs of both quantity-based schemes differ from their minimum levels by the same factor:

$$\frac{S_s^+}{S_s^*} = \frac{S_P^+}{S^*} = (k_a^{-\alpha\varepsilon} + \alpha k_a^\varepsilon)\varepsilon, \tag{4.14}$$

where S_P^+ is the social cost with permits if costs are misestimated and S^* is the social cost with either permits or charges if costs are estimated correctly.

Normalizing again by the control costs under the optimal standard, the net benefit of switching to a charge when the standard is based on incorrect cost information is given by

$$\Delta S^+ = \frac{S_s^+ - S^*}{\varepsilon S_s^*}$$

$$= (1 + \alpha)\left[(k_a^{-\alpha\varepsilon} + \alpha k_a^\varepsilon)\varepsilon - \frac{S^*}{S_s^*}\right] \qquad (4.15)$$

$$= \Delta S^* + k_a^{-\alpha\varepsilon} + \alpha k_a^\varepsilon - (1 + \alpha).$$

Shifting from the standard to permits when both are based on incorrect cost information yields the following change in efficiency:

$$\Delta S_P^+ = \frac{S_s^+ - S_P^+}{\varepsilon S_s^*}$$

$$= \frac{(k_a^{-\alpha\varepsilon} + \alpha k_a^\varepsilon)\varepsilon(S_s^* - S^*)}{\varepsilon S_s^*} \qquad (4.16)$$

$$= (k_a^{-\alpha\varepsilon} + \alpha k_a^\varepsilon)\varepsilon(\Delta S^*).$$

Figure 4.5 plots ΔS^+ and ΔS_P^+ as functions of k_a for $\sigma_a^2 = 2$ and $\alpha = 1$ (the solid lines) and $\alpha = 0.5$ (the dotted lines). If $\alpha = 1$ and the costs are estimated correctly ($k_a = 1$), the charge and permits both yield a gain of 44 percent over the standard, as before. As k_a deviates from unity, however, the efficiency advantage of marketable permits rises much less rapidly than that of the charge; if costs are misestimated by a factor of two ($k_a = 0.5$ or $k_a = 2$), for example, the efficiency gain under the charge is 56 percent, whereas permits yield a gain of only 47 percent. If $k_a = 5$ (or 0.2), the efficiency gains are 113 percent and 59 percent, respectively. The results for $\alpha = 0.5$ are similar, though the efficiency advantages of the two incentive schemes relative to the standard are smaller.

Dynamic Aspects

The basic framework outlined in the previous section also may be used to assess the relative dynamic efficiencies of charges and marketable permits. Over time, cost and damage functions shift, often substantially, due in large part to exogenous factors, though the regulatory policy itself may

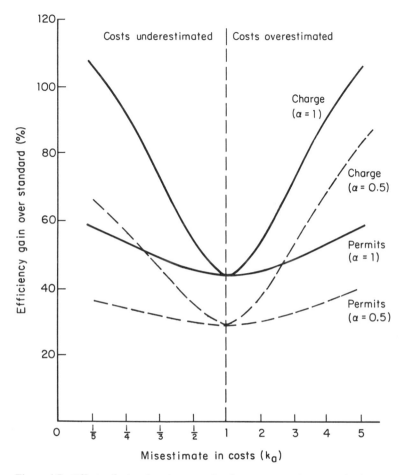

Figure 4.5   Effects of misestimating costs for charge vs. permit vs. standard

have a significant impact. Ideally, the policy should be structured both to remain optimal in the face of such shifts and to encourage beneficial changes in costs and benefits.

Accommodating change

Perhaps the most obvious strategy for an agency concerned about change is to review its regulations frequently, making modifications as necessary. As many observers have pointed out, however, regulations tend to be "sticky," changing only at irregular, often lengthy, intervals.[16] Moreover such reviews are expensive and time-consuming. At the other extreme we might assume that regulations will remain fixed for an indefinite time and

Table 4.1 Effects of changes in costs and benefits

| Nature of change | Effect | Level of emissions relative to optimum[a] | |
		Charge	Permits
General inflation	MC and MC curves shift up by equal proportions	+	0
Growth in production	MC curve shifts up	+(0)[b]	−
Innovation in control technology	MC curve shifts down	−(0)[b]	+
Growth in affected population	MD curve shifts up	+	+
Increased willingness to pay	MD curve shifts up	+	+
Improved medical treatment	MD curve shifts down	−	−

a. " + " means emissions higher than optimal (too little control); " − " means emissions lower than optimal (too much control); "0" means emissions optimal.
b. If MD is constant, shifts in MC do not affect the optimal level of the charge.

then ask in that context how well the different instruments will adjust to likely changes.

Let us consider six possible types of changes: (1) general price inflation, (2) increases in the number of emitters as an accompaniment to economic growth, (3) cost-reducing innovation in control technologies, (4) growth in the population affected by emissions, (5) increased willingness to pay for health and environmental amenities as real income rises, and (6) improvements in medical treatments that reduce the consequences of illnesses caused by emissions (e.g., the cure rate for lung cancer increases). The first change results in equal proportional upward shifts in both the marginal damage and the marginal cost curves. The second item raises the marginal cost curve, whereas the third generally lowers it. The fourth and fifth both raise the marginal damage curve. Finally, the sixth change lowers the marginal damage curve.

Table 4.1 summarizes the effects of each of the six changes and indicates whether the level of emissions will be higher or lower than optimal under either charges or permits. The effect of uniform inflation, as noted earlier, is to shift both curves upward by the same proportion; hence the optimal quantity of emissions remains unchanged, but the optimal charge rises. Thus the initial charge leads to excessive emissions. Note, however, that this problem can be dealt with easily by tying the charge to an appropriate price index. Given recent high rates of inflation, such indexing is essential if the real charge rate is not to fall rapidly.

The second change, growth in the number of emitters (or expansion in the production of existing emitters), shifts up the marginal cost curve, so the optimal level of emissions is higher than that allowed under the permit scheme. If marginal damages are rising, the charge leads to excessive emissions. If marginal damages are constant, however, the original charge rate continues to be optimal. The effects are simply reversed if innovation lowers the cost of control; emissions are too high with permits and too low with the charge, unless marginal damages are constant, in which case the charge leads to the optimal level of emissions.

As in the analysis of uncertainty, changes in marginal damage affect both schemes equally. If marginal damage rises, due either to growth in the affected population (and hence an increase in physical damages) or to an increase in the monetary valuation of damage, both schemes result in excessive emissions relative to the new optimum. If marginal damage decreases, say through improved treatment, both schemes lead to too few emissions. If marginal damage is independent of the level of emissions, however, it may be possible to structure the charge scheme so as to provide automatic, optimal adjustments for at least some changes in marginal damage. The charge, for example, can be stated as a function of the number of people living in the area affected by emissions, so that as population changes, the charge will change as well. (The exposure charges discussed in later chapters have this property.) Similarly, the charge might be indexed for changes in real per capita income as a proxy for changes in willingness to pay.

Note that similar adjustments cannot be incorporated easily into a permits scheme, because the change in the optimal number of permits depends on the marginal cost function as well as on the shift in the damage function. A 10 percent rise in population, for example, does not necessarily call for a 10 percent reduction in total emissions.[17] Similar problems arise with charges if marginal damage depends on the level of emissions. If marginal damage rises with emissions, then a shift in the marginal damage curve does not result in an equal shift in the optimal charge rate. Figure 4.6 illustrates this problem. Marginal damage shifts up by ΔD, from MD_0 to MD_1. Note, however, that the optimal charge shifts up by less, so that if the charge rate were raised automatically by ΔD, the resulting level of emissions (X_1^t) would be lower than the new optimum (X_1^*).

Stimulating Innovation

Many of the types of changes discussed here are almost entirely independent of the regulatory system. The rate of general inflation and advances in medical treatment, for example, are likely to be the same whether EPA uses charges, permits, or standards. Some kinds of changes, however,

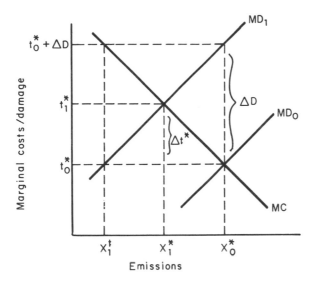

Figure 4.6   Shift in optimal charge less than marginal damage if marginal damage rising

may depend significantly on the regulatory instrument. Technological innovation in control techniques is likely to be particularly sensitive.

Under a standard firms have an incentive to develop new control techniques that reduce the cost of achieving the standard. They have no incentive, however, to develop techniques that will achieve higher levels of control (lower emissions) because they receive no "credit" for doing better than the standard requires.[18]

Under charges, permits, or subsidies, each firm has a strong incentive to innovate and to develop control techniques that will reach tighter levels of control, because under those systems a firm receives credit for further reductions in emissions. Under charges, reduced emissions mean lower payments. Under a permit system, reduced emissions mean that the firm does not have to purchase as many permits or can sell some that it already holds. Under a subsidy system, reduced emissions mean a larger subsidy. Thus firms have a stronger incentive to innovate and to apply those innovations to reduce emissions over time. Note, however, that under a permit scheme the new control techniques are not translated into a lower aggregate level of emissions; those firms that innovate may lower their emissions, but since the total level of emissions remains fixed, other firms will increase emissions as the price of permits falls.

If the charge (or subsidy) rate is held fixed, the incentive for innovation is maintained over time. Under a permit scheme, however, the incentive

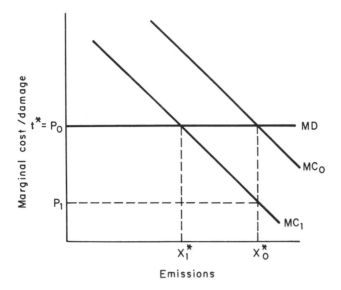

Figure 4.7 Charges maintain the incentive to innovate

to innovate tends to erode because cost-reducing innovations lower the price of the permits. Figure 4.7 illustrates the situation for constant marginal damages. In that case the optimal charge (or subsidy) rate is $t^* = MD$, regardless of marginal cost. Given the initial marginal cost curve, MC_0, the optimal number of permits is X_0^*, and their equilibrium price is $P_0 = t^*$. At that point the permit scheme provides the same incentive to innovate that the charge (or subsidy) scheme does. Suppose, however, that firms do innovate, and that the aggregate marginal cost curve falls to MC_1. Under a charge (or subsidy) emissions fall to X_1^*, and the incentive to innovate remains the same; continued reductions in emissions yield the same marginal benefit to firms. Under the permit scheme, however, emissions remain fixed at X_0^*, and the price of the permit falls to $P_1 < P_0 = t^*$; the incentive for further innovation falls. Thus over the long run charges provide a stronger continuing incentive for innovation than do permits.[19]

Summary

Uncertainty and change over time pose obstacles to the efficient operation of any regulatory strategy. A regulation that appears to be optimal when adopted may be totally inappropriate several years later, either because our knowledge has changed or because the costs and benefits have themselves changed. No instrument can guarantee instantaneous, perfect adjustment. Charges, however, appear to offer greater flexibility and responsiveness to uncertainty and change than do quantity-based schemes, such as permits and standards. A major argument made by opponents of charges—that charges are unworkable in the face of substantial uncertainties about costs and, more important, damages—is simply wrong. Contrary to popular wisdom, errors in estimating the damage function may well *increase* the efficiency of charges relative to standards, and they have no effect on the comparison between charges and permits. Quantity-based schemes may obscure the importance of marginal damage in choosing the right level of regulation, but they do not reduce it.

The shape of the marginal damage function is important to the choice between charges and quantity-based approaches. In the limited number of cases where thresholds or quasi thresholds exist, *and can be identified with reasonable accuracy*, permits are preferable to charges and dominate standards. If expected marginal damage varies little with the total level of emissions, however, as appears to be the case for many environmental problems (including airborne carcinogens), uncertainty about control costs increases the comparative advantage of charges over both marketable permits and standards.

Similar reasoning suggests that in most cases charges will outperform quantity-based schemes in a dynamic context as conditions change. If expected marginal damage does not vary with total emissions, a charge will continue to yield the optimal level of emissions as marginal costs change due to, for example, changes in production levels or the introduction of technological innovations. Moreover a charge will provide a stronger ongoing incentive for firms to develop new control technologies.

As under uncertainty, price and quantity-based schemes are both sensitive to shifts in the marginal damage function. Charges can be designed to adjust automatically to many types of changes in marginal damages, however. Indexing the charge to the general price level to maintain its real incentive effect is perhaps the most obvious need, but more subtle adjustments also can be included. Good cases can be made, for example, for linking the charge rate to the size of the population affected (thus capturing a major source of change in physical damages) and to real per capita personal income (an important indicator of willingness to pay for environmental improvement). Similar provisions for automatic adjustments are far more difficult to build into quantity-based strategies.

Chapter 5

An Overview of the Targeting Problem

Designers of regulation must specify not only the instrument to be employed, but also the target(s) to which it will be applied. Traditionally, theoretical analyses have taken the target as given, as we did in the three preceding chapters. They have assumed that emissions are the object of regulation, and that the benefits of different strategies can be summarized in terms of their impacts on aggregate emissions.[1] Most environmental externalities, however, are the products of complex, multistage processes that offer many potential points of intervention. A source's emission level is only one of these potential targets. In many cases other targets may be more appropriate, because they are much easier to monitor or because they lead to a more cost-effective allocation of control efforts.

The choice of target has important implications for both the ease with which a regulation can be enforced and its efficiency. What do we sacrifice in efficiency, for example, if we aim standards at the specifications of control equipment because it is difficult to measure emissions? If the marginal damage caused by a unit of emissions varies across sources, can we improve efficiency by switching the target from emissions to exposure? Under what conditions will a uniform exposure standard be more efficient than a uniform emission standard? This chapter and the next two attempt to answer these questions and, more generally, to develop guidelines for choosing among alternative targets, as well as among instruments.

These two dimensions of choice—what instrument to use and where to target it—though conceptually distinct, are by no means independent. Just as the optimal level of control depends on the instrument, as discussed in chapter 2, the optimal target may vary with the regulatory instrument. As shown formally in the next chapter, for example, emissions may be the preferred target for a standard, whereas exposure will be the better target for a charge. Similarly, the choice of instrument may depend on the targets available. The gains of switching from standards to charges may be much greater, for example, when monitoring problems force the regulator to use multiple, imperfect targets rather than a single, more comprehensive measure. Thus we need to examine strategies, combinations of instruments and targets, rather than either dimension in isolation.

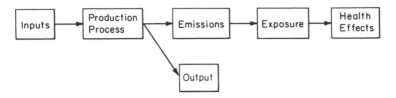

Figure 5.1 Multiple stages providing multiple targets for intervention

The Basic Framework

The ultimate goal in regulating carcinogens and other health-threatening substances is to reduce adverse health effects. The final outcome is a function of many variables, some of which are subject to control by individual sources, others of which must be regarded as fixed, if not absolutely then at least in a practical sense. Regulations may be targeted quite specifically on particular variables or small clusters of variables, or they may be based on summary measures that are functions of many or all of the variables.

Figure 5.1 represents the framework in schematic form as a series of sequential stages leading from the inputs used in the production process, through several intermediate stages, to final health effects. As drawn, the diagram is most applicable to an industrial plant emitting an environmental pollutant, but the basic principles apply to a much broader range of cases.

Inputs, Production and Emissions

The diagram begins with the inputs—labor, capital, and raw materials—used in the production process. Examples of existing input-based standards include bans imposed by EPA on the use of certain pesticides and most of OSHA's safety standards, which specify certain characteristics of plants and equipment. Inputs also can be the targets of incentives, as with the tax on sulfur in fuels proposed by the Nixon administration in the early 1970s (Anderson et al. 1977, 51–53), and the marketable permits scheme for chlorofluorocarbons studied by EPA (Rabin1981).

The second stage in the diagram is the production process. Damages often can be reduced by modifications made at this stage. Emissions from coke ovens, for example, depend on the procedures used to add coal to the ovens and on how long the coking process continues before the ovens are opened to remove the finished coke (Whiting 1979). Emissions from fuel transfer operations can be reduced by switching from "splash fill" to "submerged fill," in which the hose is lowered into the fuel already in the tank being filled (U.S. EPA 1978b). Regulations aimed at this stage include provisions of EPA's proposed generic policy toward airborne carcinogens that mandate inspection and maintenance procedures to reduce leaks from valves and pumps (44 *Fed. Reg.* 58642 1979).

The third stage is split between output and emissions. The former is the good or service produced, whereas the latter is the amount of the hazardous substance released into the environment as a by-product. In this formulation the level of output does not have a direct causal link to exposure and health effects. Reducing output, however, generally will reduce emissions. As discussed in chapter 3, until relatively recently most of the theoretical literature on externalities assumed implicitly that the output and the externality were produced in fixed proportions, in which case controlling the level of output is equivalent to controlling the level of the externality; indeed, it is the only way. Under that formulation, for example, to reduce the emissions of sulfur from electric power generating plants, the amount of power generated would have to be reduced through a tax on electricity or some other means.

In most cases the link between output and emissions is not fixed, so quantity restrictions or taxes aimed at output are not very efficient methods for controlling the associated externality. Rather than cut back electricity generation to reduce sulfur emissions, for example, utilities can install scrubbers to remove sulfur from the stack exhaust gas. Interventions at the emissions stage typically are based on performance measures. EPA's proposed standard for existing maleic anhydride plants, for example, states that benzene emissions shall not exceed 0.3 kg of benzene per 100 kg of benzene input (45 *Fed. Reg.* 26660 1980). Such regulations, although targeted at the emissions stage, are based on cumulative measures and thus affect firms' decisions at earlier stages as well. Power plants faced with a limit on sulfur emissions, for example, can use stack scrubbers, but they also can meet the standard by making changes at earlier stages, say, by switching to low-sulfur fuels or by modifying the combustion process. Some interventions targeted on emissions are not cumulative, however. A standard, for example, that specifies that coal-fired power plants must remove a certain percentage of sulfur from the exhaust gas provides little or no incentive to make modifications at either the input or the production stages.

Exposure

Discussions of regulatory alternatives usually ignore the two stages beyond emissions in figure 5.1, human exposure and health effects, or assume (as we did in earlier chapters) that the relationship between emissions and health effects is fixed and uniform across sources so that health effects will be proportional to total emissions. These two stages, however, merit consideration for at least three reasons: (1) they may offer additional options for reducing damages that will not be exercised if the regulation is targeted at an earlier stage; (2) even where the link between emissions and damages is fixed for a given source, if it varies across sources the optimal degree of control at earlier stages also will vary; and (3) in a few instances, exposure or health effects may be easier to monitor than indicators at earlier stages, such as emissions.

Human exposure simply means: how many people are exposed to what concentrations over what periods of time. In some cases exposure can be reduced without reducing emissions. In the workplace, for example, for any given concentration of a hazardous substance, aggregate exposure can be reduced by reducing the number of workers in the area. OSHA's standards limiting concentrations provide no incentives for firms to minimize the number of workers exposed.[2] Similarly, with environmental pollutants, for any level of emissions, human exposure can be reduced by shifting the location of a plant to a less densely populated area or by varying operations depending on meteorological conditions.[3] Emission standards and charges provide no incentive for firms to exercise these control options.

Interventions at the exposure stage also may have a cumulative effect, as with a charge levied on exposure, or they may be directed at that stage alone. Examples of the latter include prohibitions on the transportation of hazardous substances by truck through tunnels and densely populated cities, and a tax on coal-mine operators to finance black lung benefits based solely on the number of miners employed. Even where it is not possible to alter the emissions-exposure link, it may be desirable to target the intervention on exposure rather than emissions. Service stations, for example, have relatively little flexibility in choosing their locations; they must locate where their potential customers are. Yet each kilogram of benzene emitted from an urban station causes far more exposure than one from a rural station so that it makes little sense to treat emissions from the two types of stations as equivalent for regulatory purposes.

Health Effects
The final stage in the process is the adverse health effects, such as illness or death, that may result from exposure. The relationship between exposure and health effects depends on the dose-response relationship(s) for

the individuals exposed and on the efficacy of medical treatments. The opportunities for damage-reduction strategies at this stage are likely to be very limited, if not nonexistent, for environmental hazards. The occupational setting may provide greater opportunities. For a given level of exposure, a plant can lower risk to its employees by screening out workers most susceptible to the hazard in question. Some asbestos producers, for example, will not hire smokers because of the synergistic effects of smoking and asbestos exposure in inducing lung cancer. Similarly, some companies will not allow women of childbearing age to work at jobs where they will be exposed to lead, a teratogen. As scientific understanding of the factors that enhance risk increases, this type of screening is likely to grow in frequency and sophistication.

Firms also may try to reduce the adverse effects of exposure to their employees through improved medical treatment. OSHA's 1978 benzene standard (43 *Fed. Reg.* 5918 1978), for example, requires regular medical examinations of workers exposed to benzene, the apparent rationale being that such examinations may detect hematological changes believed to be precursors of leukemia, so that workers with such symptoms could be removed from jobs exposing them to benzene or at least have their leukemia diagnosed early so that treatment could begin.

These kinds of options for altering the link between exposure and health effects are unlikely to be significant for environmental hazards because the levels of exposure and risk involved typically are much lower than in the workplace. It also seems unlikely that the link between exposure and risk will vary widely across sources, at least in measurable or predictable ways. Thus little is likely to be gained by shifting the target of environmental regulation from exposure to actual health effects. Moreover, even if it were desirable to do so, it would rarely be possible. In most cases individual health outcomes cannot be tied to specific substances, let alone to specific sources. We have evidence, for example, that exposure to benzene raises the risk of leukemia, but there is no way to show that any particular case of leukemia was caused by benzene exposure. Even if there were, we could not say which of the many sources of benzene emissions was responsible.

Where it is possible to link specific cases of disease to certain pollutants, but not to individual sources, one alternative is to compensate victims from funds collected from the polluters as a group. Japan uses such a system for sulfur oxides, but the fees paid by individual sources are based on weight for automobiles and on emissions for industrial plants, so it cannot truly be classified as a health-effects charge (Anderson et al. 1977,

50). Its incentive effect on industrial polluters is the same as that of an emission charge; the effect on automobile owners is unclear.

Criteria for Choosing the Target

The preceding discussion suggests the wide range of potential targets, though by no means does it exhaust all of the alternatives. In particular, the possibility of combinations of several targets for any given source category greatly expands the number of strategies available.

Efficiency Conditions

To achieve full efficiency, a regulatory strategy must (1) allocate control efforts *within* each source to minimize the cost of achieving any given reduction in damages from that source, (2) allocate control efforts *across* sources in a manner that minimizes the cost of achieving any given reduction in overall damages, and (3) strike an appropriate balance between the costs of control and the benefits of damage reduction.

Achievement of the third condition depends primarily on the judgment of decision makers and the quality of the data available. As shown in chapter 4, however, charges are more likely than quantity-based approaches to maintain the desired trade-off between control costs and damages under conditions of uncertainty. Moreover the choices of targets and instruments may affect the clarity with which the trade-offs are presented to decision makers and the ease with which they can make comparisons across different regulations. For example, EPA would be more likely to ensure consistent trade-offs between cost and health protection in regulating different categories of benzene emission sources by using an exposure charge (the charge rate would be the same for all categories), than by setting an emission standard for each category (the optimal standard for a category would depend on its emission control costs and the link between emissions and damage).

The second optimality condition has been the central focus of the literature on alternative instruments; the primary virtue of incentive schemes, as shown in chapter 2, is that they allocate control levels efficiently across sources. If the target is chosen incorrectly, however, a charge may be far from optimal. A uniform emission charge, for example, does an excellent job of allocating emissions across sources, but, as shown in the next chapter, it may do a very poor job of allocating damage reductions.

Achievement of the first optimality condition, efficient allocation within sources, is primarily a function of the target. Note that the traditional formulation assumes away the problem; if the damage from each source is solely a function of its emission level and the target of regulation is emissions, then cost-minimizing behavior on the part of firms will ensure that each source will select the least-cost mix of control variables for achieving its reduction in damages, regardless of the regulatory instrument. If emissions are not the target of regulation, however, or if damages also depend on other source-specific variables, then we cannot be certain that the condition will be satisfied. Improperly designed targets may lead firms to concentrate on certain kinds of damage-reduction activities when other, more cost-effective means are available.

Efficient Targets for Charges

All three efficiency conditions can be achieved by levying a charge directly on the damages caused by each source. Under such a charge sources have maximum flexibility in determining the mix of control efforts they undertake, thus allowing them to choose the most efficient combination for their individual circumstances. The charge also allocates control efforts efficiently across sources. The occupational injury tax proposed by Smith (1976) is an example of this type of damage-based charge. Under it firms would have an incentive to exercise all of the many options they have available to reduce threats to workers' safety, including training, supervisory practices, and other factors not addressed effectively by current OSHA standards. An injury tax also would equalize the marginal costs of reducing risks across firms. Moreover it almost certainly would be easier to monitor and enforce than present regulations, particularly if it were linked to existing workers' compensation systems.

In most cases, unfortunately, it is impossible to target a charge directly on damages. We have no way, for example, to link specific cases of leukemia with the benzene released from particular maleic anhydride plants. It may be possible, however, to devise targets that closely approximate the effects of a damage charge. Exposure is often an excellent proxy for environmental health hazards because, as discussed earlier, the link between exposure and health effects is likely to be quite uniform across sources and not subject to alteration by firms. Indeed, even if it is possible to measure actual damages directly, exposure may be a better target if firms are risk averse; the level of payments under a charge levied on actual damages will be highly uncertain because of the stochastic link between exposure and damage.

What do we lose as we move the target of the charge farther away from damages? One obvious possibility is that the charge may no longer have any impact on some control variables that sources could alter to reduce damages. As suggested earlier, for example, a uniform emission charge provides no incentive for firms to locate hazardous activities where exposure will be low. The farther the target is removed from the outcome measure of ultimate concern, the more likely it is that cost-effective opportunities for control will be missed. A charge levied on the production of benzene, for example, would reduce exposure by reducing the amount of benzene used. It would do nothing, however, to encourage firms to install emission control devices, except insofar as the devices captured benzene for reuse.[4]

For cases where it is difficult to devise a single target that encompasses all control variables simultaneously, one option may be to impose separate charges on several different targets. If the contributions of these targets to damages are independent of one another, this strategy will not entail any loss in efficiency. Refineries, for example, emit benzene from many different sources, including storage tanks, valves, and various processing units. The effect will be the same whether a charge is levied on benzene emissions from each source or on the total of emissions from all sources.

If the control variables have interdependent effects on damages, however, separate charges generally will not be efficient. The benzene emissions from a service station, for example, are the product of the benzene content of the gasoline and the efficiency of the evaporative controls (if any) installed at that station. As it is extremely difficult to measure benzene emissions from service stations, one might be tempted to impose one charge on refiners based on the benzene content of gasoline and another on individual service stations based on the efficiency of the controls in place. Such a two-charge strategy would equalize across refineries the marginal costs of reducing the benzene content of gasoline and would equalize across service stations the marginal costs of vapor recovery. Note, however, that the marginal benefit of reducing the benzene content of gasoline at a particular refinery depends on the evaporative controls at the service stations where the gasoline is sold; the higher the level of controls, the lower the gain from reducing benzene content. Similarly, the lower the benzene content of the gasoline, the lower the payoff from installing evaporative controls. Thus, as shown formally in chapter 7, a single cumulative target for a charge will be more efficient than multiple individual targets.

Choosing a target well removed from damages may entail a significant loss in efficiency even if the target encompasses all of the control options open to firms. The reasoning is similar to the argument made earlier. A uniform charge equalizes across sources the marginal costs of controlling the targeted measure. Thus, for example, a uniform emission charge equalizes across firms the marginal costs of controlling *emissions*. Efficiency, however, requires that the marginal costs of reducing *damages* be equalized across sources. If the link between the target and marginal damages varies, a uniform charge cannot achieve this optimum, even if firms have no control over the link. This issue is explored at greater length in the next chapter.

Efficient Targets for Standards
As we have seen, the ideal target for a charge is the external damage itself. Although this is rarely possible, it suggests that, if the instrument is a charge, the target selected should be cumulative in nature and as close as possible to the outcome measure of ultimate concern. It is tempting, but erroneous, to reason by analogy that similar advice should be given to a regulator choosing the target for a standard.

One difficulty is that standards cannot be imposed on targets that exhibit substantial random fluctuations. It would not be possible, for example, to impose an occupational injury standard analogous to the injury tax discussed earlier because the number of workers actually injured at any given plant over any time period is stochastic. A plant that employs 100 workers, each subject to an annual risk of injury of 0.01, for example, will experience on average one injury per year. But the actual number injured in any one year may vary considerably. Even if the risks to workers are independent, in any given year there is a 0.37 chance that no workers will be injured, and a 0.26 chance that two or more will be injured.[5] If injuries are not independent (e.g., there are multi-injury accidents), which is likely to be the case, the variation in the number of injuries will be greater still. This variation causes few, if any, problems under a charge system, other than the possible risk-spreading losses mentioned earlier, but it renders unworkable a fixed standard. One solution would be to set a standard and then impose a modest fine on injuries in excess of the standard. At that point, however, the strategy begins to resemble a charge more than a standard.[6]

Even if the most efficient target for a charge does not have a random component, it may not be optimal for a standard. An all-inclusive target leads to an efficient allocation of control efforts *within* sources, whether the instrument is a charge or a standard. With a standard, however, it may lead to a grossly inefficient allocation of control efforts *across* sources. Moving the target of a uniform standard closer to damages may or may not improve matters, as shown in the next chapter when the target is switched from emissions to exposure.

Consolidating separate standards into a single target, however, generally will increase efficiency. EPA's bubble policy is an example of such a change. Under existing regulations a single plant may have many different sources emitting the same pollutant, each subject to a separate standard. The bubble policy allows a plant to exceed the standards for some sources if it makes offsetting reductions in emissions from other sources (Clark 1979). In some sense it allows each plant to set up an internal market in emission permits. The change increases efficiency because it yields the same level of emissions (from each site) as current source-by-source standards but allows firms more flexibility in choosing the least-cost mix of control efforts.

An Example
Efforts to reduce gasoline consumption through regulation illustrate the interactions between the choices of targets and instruments. One approach would be to set a uniform mileage standard that would apply to all automobiles. Each car sold, for example, might have to average at least 25 miles per gallon. It is much more expensive to achieve higher mileage with large cars than with small cars, however, so a uniform standard would be highly inefficient. The current regulatory strategy, the "fleet mileage standard," largely avoids that problem by making the standard apply not to individual automobiles but rather to the sales-weighted average of all automobiles sold by each manufacturer. The system is very similar to the bubble concept and has similar efficiency advantages over a standard targeted on individual sources (automobiles).

Efficiency could be increased by switching to a "gas-guzzler" tax or to a system of marketable permits, both of which would permit trade-offs across companies as well as within them. The gain, however, probably would be farily modest because all of the major U.S manufacturers produce a mix of both small and large cars; the gain would be larger if some manufacturers would find it advantageous to specialize in large cars, while others would concentrate on producing smaller cars.

All of the strategies discussed thus far are targeted on the fuel efficiency of new cars. Gasoline consumption, however, is also a function of many other variables, including maintenance, number of miles driven, and driving style (speed, rate of acceleration, etc.). The fleet mileage standard has no systematic effect on any of these variables, nor would a gas-guzzler tax. Moreover both approaches in essence contain a "grandfather clause," as they apply only to automobiles produced after the regulation is promulgated.

The obvious alternative is to impose a tax (charge) directly on gasoline, thus providing individuals with an incentive to adjust all of the variables under their control that affect the consumption of gasoline. Faced with a tax on gasoline, people presumably not only would purchase more fuel-efficient cars but also would drive fewer miles at lower speeds and keep their cars in better tune. Owners of older, less fuel-efficient cars would have an incentive to junk them sooner.

The tax on gasoline could yield significant efficiency gains even if it had little impact on the drivers' behaviors after they purchased cars. Suppose, for the sake of argument, that individuals have little or no control over the number of miles they drive. That number, however, varies widely, from some traveling salespersons who drive 50,000 miles per year, down to some elderly people who drive only 1,000 miles per year. Clearly, the return in terms of reduced gasoline consumption to investment in greater fuel efficiency is far higher in the former case than in the latter. Under the current strategy or under a gas-guzzler tax, however, both types of drivers face the same incentive to buy more fuel-efficient cars. In contrast, a tax levied on gasoline provides the appropriate differential incentives, giving extra encouragement to the salesperson to buy a more fuel-efficient car. A system of transferable gasoline ration coupons (marketable permits) would have essentially the same impact as a tax on gasoline.

We also could target a standard on gasoline consumption; each auto-mobile owner could be limited to some specified number of gallons of gasoline a year. In essence that would be a nontransferable rationing system. Just as with a tax on gasoline or a transferable rationing system, such a plan would induce each car owner to choose the most efficient mix of actions available to him to reduce gasoline consumption. The allocation across individuals, however, would be highly inefficient. For some the standard would not be binding; the marginal cost would be zero. For others such as our hypothetical salesperson, however, the marginal cost would be extremely high. Thus switching the target of the standard probably would decrease rather than increase efficiency, despite the fact that the current fleet mileage standard affects only one of several variables that determine gasoline consumption.

Ability to Monitor

Thus far, although we have alluded to the monitoring problem, we have not dealt with it directly. Regardless of the instrument employed, the regulator must be able to measure the target selected, either directly or indirectly, if the regulation is to be enforceable. Regulators have long complained that economists pay insufficient attention to this important practical issue. A common criticism of emission charges is that for many sources it is exceedingly expensive, if not impossible, to monitor emission levels accurately on a continuous basis. In contrast, standards that require the use of particular inputs or the installation of specific control devices may be far easier to enforce.

Ease of monitoring, however, is predominantly a function of the target, not the instrument. Both emission standards and emission charges, for example, require that the regulator be able to monitor emission levels. In contrast, monitoring problems may vary widely across targets for a given instrument. It would be relatively easy, for example, to monitor compliance with a standard specifying the design of benzene storage tanks but relatively difficult to determine if those tanks complied over time with an emission standard.

For a given target, compliance may be slightly easier to monitor with standards than with incentives because the former require only a yes or no determination, whereas the latter require the ability to measure a wider range of outcomes. The distinction in automobiles between warning lights and gauges for detecting problems with engine temperature, oil pressure, and the like, offers a simple analogy; the warning light is simpler and cheaper because it only needs to distinguish between two states of the world. Similarly, the monitoring device needed to tell whether a standard has been exceeded may be cheaper than one that must record varying emission levels over time to determine what charge should be assessed. It seems unlikely, however, that the cost differential would be significant in most cases.

Accurate monitoring may be *more* important with a standard, at least over a narrow range. To support enforcement action, the regulator must have good evidence that the standard has been violated, that the high reading is not the result of errors in monitoring. In contrast, under a charge system errors in monitoring are less critical, because the penalty function is continuous; random fluctuations in measurements lead to relatively modest variations in the charge paid, variations that are likely to average out over time.

Where it is not possible to monitor a particular target directly, indirect means often can be used. Individual exposures, for example, cannot be measured on an ongoing basis. As discussed more fully in the next chapter, however, the regulatory agency can estimate exposure factors for each source or, where there are many sources, for groups of sources in a given geographical area, using dispersion models and data on meteorological conditions and population patterns. These exposure factors then can be used in conjunction with measured emissions to estimate exposure on an ongoing basis.

Other targets also may be monitored indirectly. Where emissions cannot be measured on a continuous basis, for example, the regulator has several alternatives. One is to take random samples of emissions and to use those results to estimate the total. Another is to measure emissions under various operating conditions and then use more easily measured operating parameters, such as production volume or the amount of electricity consumed by control devices, to estimate emissions over time. Such indirect measures inevitably will be imperfect, but they may well be the best choice if the alternatives are to engage in far more expensive forms of monitoring or to select a far less appropriate, but directly measurable, target.

Summary

The choice of target plays a dominant role in determining the ease with which a regulation can be enforced. Together with the choice of instrument, it also plays an important role in determining how efficiently control efforts are allocated both within and across sources. Targets and instruments must be evaluated simultaneously; in general, we cannot rank alternative targets and alternative instruments independently.

If the instrument is a charge, the goal is to select a target that is as close to external damages as monitoring capabilities and administrative feasibility permit. The ideal target is the outcome measure itself, though other targets may be almost as efficient and much easier to monitor.

If the instrument is a standard, the optimal target is much less clear. As with a charge the goal of efficient allocation of control efforts within sources suggests the desirability of choosing a cumulative, all-inclusive target. With a charge such a target also contributes to the efficient allocation of control efforts across sources. With a standard, however, a more comprehensive target may or may not improve overall efficiency. Moreover targets with a stochastic component may not even be feasible for a standard, as illustrated by the earlier comparison between an occupational injury charge and an occupational injury standard.

This chapter has attempted to present a class of important issues that has received relatively little attention in the literature. Once we relax some of the assumptions made in earlier chapters, the design of regulatory strategies is seen to be a problem of the "second best."[7] As, in general, with second-best problems we cannot simply rank alternatives on the basis of how many optimality conditions they satisfy. A related characteristic of second-best problems is that the answers often are neither crisp nor definite on a general level; the best solution in a second-best situation typically depends on the specifics of the individual case. The next two chapters return to a more formal mode of analysis, using some relatively simple mathematical models to obtain more precise results for special cases of the targeting problem.

Chapter 6

Variability in Marginal Damages: Emissions vs. Exposure Targets

Uniform emission standards suffer from two major faults. First, they fail to take account of variations across sources in the cost of controlling emissions. That fault has received most of the attention of economists and serves as the primary basis for recommending uniform emission charges or marketable emission permit schemes. Second, uniform standards fail to take account of variations in the marginal damages caused by emissions. Even if two sources have identical control costs, a uniform emission limit will not be efficient if the damages caused by emissions differ between the two plants. This issue has received relatively little attention in the literature, though it is often very important in practice.

Conventional emission charges and permits also ignore this variability in marginal damages. Under a uniform emission charge a firm pays the same per unit emitted whether 10 or 10,000 people will be exposed as a result, though clearly the emissions cause more damage in the latter case. Marketable emission permits suffer from the same problem; permits can be traded freely among sources with radically different marginal damages.

Variability in marginal damages is not an academic curiosity. As shown in chapter 9, for example, a unit of benzene emitted from one maleic anhydride plant may cause almost 50 times as much damage as a unit emitted from another plant. The marginal damages associated with benzene emissions from service stations and automobiles also may vary widely, by two orders of magnitude or more.

As discussed in chapter 5, the marginal damages caused by emissions are likely to vary almost exclusively because exposure patterns vary, due to differences in meteorological conditions and, more important, in population densities. We have no reason to suspect, for example, that residents of New York City are more sensitive than residents of rural Nevada to benzene-induced leukemia, but we do know that there are far more New Yorkers to be exposed. Thus exposure measures offer reasonable proxies for damages.

Previous Studies

The issue of variability across sources in the link between emissions and damages has received only a limited amount of attention in the literature. Rose-Ackerman (1974), for example, in an article generally critical of effluent charges, points out that if marginal damages vary across sites, a simple uniform effluent charge will not be optimal and that "it is only a sophisticated effluent charge which is certain to be more efficient than a primitive nonmarket mode of allocation." She stresses the difficulty of designing an optimal charge scheme when the regulatory agency is concerned about water quality at many points along a river, and the quality at each point is differentially affected by discharges from different sources.

Other authors have addressed much the same problem in the context of using charges to meet an ambient standard. Atkinson and Lewis (1974) note that uniform emission charges lead to a least-cost solution for controlling a given amount of emissions, but not for achieving a given level of ambient air quality. Under a uniform charge the charge is raised until the concentration in the most heavily polluted sector reaches the required level. The "problem" is that all of the other sectors will have cleaner air than necessary; some sources could reduce control efforts, and costs, without violating the ambient standard. Using data from St. Louis, Atkinson and Lewis find that, although an emissions least-cost strategy of the type achieved through uniform charges is considerably cheaper than one based on a uniform standard, as reported in chapter 2, an ambient least-cost strategy yields even larger savings.[1]

The Delaware Estuary study discussed in chapter 2 found similar results; while a uniform effluent charge could achieve any of the ambient standards at lower cost than a standard requiring uniform percentage reductions in discharges, it was more costly than the least-cost solution found by linear programming. That study also found that, by dividing the estuary into three zones and levying a different charge on effluents in each zone, the least-cost solution could be approximated very closely.[2]

Differences across firms in marginal damages also have implications for the design of marketable permit schemes. Montgomery (1972) analyzes two types of permits. One, a market in "emission licenses," is the usual approach whereby firms trade authorizations to emit a particular substance. The other, a market in "pollution licenses," establishes permits in air quality rather than emissions. Under that scheme a source might have to purchase licenses in several different markets because its emissions would affect air quality in several areas. Montgomery concludes "that the market in pollution licenses will be more widely applicable than the market in emission licenses." Tietenberg (1974a, b) proposes a similar approach, an "air-rights market." As under Montgomery's "pollution licenses," each source probably would have to purchase rights or permits in more than one market.

Tietenberg's study and the others remind us of the importance of the spatial component and emphasize that the total reduction in emissions is not a sufficient statistic for comparing regulatory strategies. The alternative criterion, however, achieving an ambient standard at least cost, has its own serious shortcomings. An ambient least-cost strategy, as Atkinson and Lewis refer to it, is not likely to be a damage least-cost strategy, one that achieves a given reduction in damages at minimum cost.

Under the ambient least-cost criterion, a strategy that just satisfies the constraint at every point is treated as equivalent to one that achieves higher air quality at all but the worst location; it gives no "credit" for doing better than the ambient standard, though in fact reductions in concentrations below the standard are likely to provide some benefit. Treating the ambient standard as an exogenous variable, set perhaps by the legislature, is a convenient way of sidestepping the very difficult problem of estimating a damage function. In many cases such ambient standards may be binding legal constraints. It is an error, however, to treat all strategies that meet the same ambient standard as if they were identical in terms of damages, differing only in their costs of control.

The General Model

Let us reformulate the problem, more appropriately, as one of minimizing the sum of control costs and damages, while recognizing that damages will depend on the distribution of emissions across sources and not just on the sum of emissions. The solution that results rarely will correspond to either the traditional emissions least-cost strategy or the ambient least-cost solutions discussed in the preceding section.

As in earlier chapters, let the ith source's emission level be denoted x_i. The cost of control for the source is $C^i(x_i)$, where $C^i(\,\cdot\,)$ varies across firms, but all experience positive and increasing marginal control costs $(-C_x^i > 0;\ C_{xx}^i > 0)$. Let the ith source's contribution to damages be $D^i(x_i)$, where $D^i(\,\cdot\,)$ varies across sources due to differences in exposure patterns and, possibly, other factors. As always, we assume that marginal damages are positive and nondecreasing $(D_x^i > 0,\ D_{xx}^i \geq 0)$. Note that under this formulation the damages caused by a source's emissions are independent of the emission levels of other sources; this assumption makes sense if expected damages are a linear function of emissions from individual sources, or if sources are so sparsely distributed that no individual is affected by emissions from more than one source. The former interpretation seems more reasonable. Total social cost is then given by

$$S = \sum_{i=1}^{n} C^i(x_i) + \sum_{i=1}^{n} D^i(x_i). \tag{6.1}$$

Differentiating with respect to each x_i yields the first-order conditions for minimum social cost:

$$-C_x^i = D_x^i, \quad i = 1, \ldots, n, \tag{6.2}$$

which simply states that for each source the marginal cost of controlling emissions should be equal to the marginal damage caused by its emissions. Note, however, that this condition is not equivalent to the more usual one that the marginal cost of controlling emissions be constant across sources. Thus, in general, $-C_x^i \neq -C_x^j$ at the optimum, and no simple uniform emission charge will allocate control efforts efficiently. The optimum requires instead that the marginal cost of controlling *damages* be constant across firms, as a simple rewriting of equation (6.2) confirms:

$$-\frac{C_x^i}{D_x^i} = 1 = -\frac{C_x^j}{D_x^j}, \quad i, j = 1, \ldots, n. \tag{6.3}$$

As this form of the optimality condition makes clear, a uniform charge levied on *damages* allocates control efforts efficiently. Alternatively, the charge may be levied on some variable that is proportional to damages, such as exposure in the case of a substance with a linear dose-response function.

Targeting Standards on Damages
Variability in the relationship between emissions and damages across firms suggests that standards also might be based on damages or exposure rather than emissions. In setting standards for occupational exposure to radiation in nuclear power plants, for example, the Nuclear Regulatory Commission (NRC) might specify that total exposure to all workers in a plant shall not exceed R person-rems (a measure of radiation exposure) per year, rather than specifying the allowable exposure per worker.[3] The NRC might take a similar approach to limiting radiation exposure for the general population so that, for example, if two plants were identical in every regard except that one was located in an area twice as densely populated as the other, it would in effect have to reduce its emissions to half the level permitted at the second plant. A less extreme approach

would be to set an emission standard that varied with population densities, though not so much as to achieve uniform levels of exposure. The "two-car" strategy advocated by some analysts in the early 1970s when automobile emission standards were being set is a crude example of that approach. Under the two-car strategy, all cars in a given area would have to meet the same emission standard, but the standard would be tighter in urban (high-exposure) areas than in rural (low-exposure) areas. As Harrison (1975) shows, compared to the uniform strategy adopted, such a plan would have cut costs dramatically with only a modest reduction in benefits.

This general model indicates that, to achieve a fully efficient allocation of control efforts, the charge should be based on damages or exposure rather than on emissions. To differentiate among the various suboptimal strategies and to gain some insight into the factors that affect the relative social costs of the alternatives, however, we need to specify the structures of the cost and damage functions in more detail. In moving from the general to the specific, we proceed in two steps, first specifying the functional forms of costs and damages, and then specifying how the parameters of these functions are distributed across sources.

Specific Cost and Damage Functions

Let the elasticity of control costs with respect to emission levels be constant across sources, as in chapter 2: $C^i(x_i) = a_i x_i^{-\alpha}$, where a_i is fixed for any particular firm, though it varies across firms. Also in keeping with earlier analyses, it is useful to interpret the x_i's as emissions per unit of production.

Let θ_i represent the ith source's exposure factor, the level of exposure per unit of substance emitted. In the analysis of maleic anhydride plants in chapter 9, for example, the exposure factors are measured in terms of ppb-person-years for each kilogram of benzene emitted, where ppb stands for part per billion and is a measure of the concentration. The product $x_i \theta_i$ is thus the total level of exposure from the ith source. If expected damages are proportional to exposure, damage from the ith firm is given by

$$D^i(x_i) = \lambda x_i \theta_i, \quad i = 1, \ldots, n, \tag{6.4}$$

where λ is the shadow price on exposure, the risk per unit of exposure times the monetary value placed on reducing risk. Total damage is simply the sum over all sources of equation (6.4).

The Options

Let us consider two alternative instruments—standards and charges—and three alternative targets. The conventional target for both instruments is emissions. A second possibility is to continue to target the intervention on emissions but to make the level of the standard or charge conditional on the exposure factor. The third alternative is to direct the instrument at exposure. Given the assumption that damage is proportional to total exposure, exposure and damage represent equivalent targets.

The combination of two instruments and three targets yields six potential regulatory approaches: (1) a conventional *emission standard* that sets a uniform value for x_i across all firms (e.g., no source may emit more than $\bar{x}$ grams of benzene per kilogram of maleic anhydride produced); (2) a *conditional standard* that sets a standard on x_i conditional on the value of θ_i (e.g., maleic anhydride plants in densely populated areas must meet a tighter emission limit than those in lightly populated areas); (3) an *exposure standard* that fixes a uniform limit on exposure, $x_i\theta_i$, from each unit of production (e.g., no plant shall cause more than $\bar{d}$ ppb-person-years of exposure to benzene per kilogram of maleic anhydride produced); (4) a conventional *emission charge* that imposes a uniform charge on x_i (e.g., each firm shall pay a charge on emissions at the rate t_x per kilogram of benzene emitted); (5) a *conditional charge* that sets a charge on emissions conditional on the value of θ_i (e.g., the higher the exposure factor, the higher the charge per unit of emissions); and (6) an *exposure charge* that imposes a uniform charge on $x_i\theta_i$ (e.g., each maleic anhydride plant must pay a charge of t_d per estimated ppb-person-year of exposure to benzene resulting from its operations). As will be demonstrated, the optimal conditional charge and the exposure charge yield identical results, leaving five distinct alternatives.

Derivations

The goal under each regulatory regime is to minimize social cost:

$$S = \sum_{i=1}^{n} a_i x_i^{-\alpha} + \lambda \sum_{i=1}^{n} x_i\theta_i. \tag{6.5}$$

Differentiating with respect to each x_i yields the first-order conditions for an optimum:

$$\frac{\partial S}{\partial x_i} = -\alpha a_i x_i^{-\alpha-1} + \lambda\theta_i = 0, \quad i = 1, \ldots, n. \tag{6.6}$$

Solving for each x_i, we obtain

$$x_i^* = \left[\frac{\alpha a_i}{(\lambda \theta_i)}\right]^\varepsilon, \qquad (6.7)$$

where, as in earlier chapters, $\varepsilon \equiv 1/(\alpha + 1)$ may be interpreted as the own-price elasticity of demand for the right to emit. Substituting equation (6.7) in equation (6.5) yields an expression for minimum social cost:

$$S^* = \left(\frac{\lambda}{\alpha}\right)^{\alpha\varepsilon} (1 + \alpha) \left[\sum_{i=1}^{n} a_i^\varepsilon \theta_i^{\alpha\varepsilon}\right] = \left[\left(\frac{\lambda}{\alpha}\right)^{\alpha\varepsilon} (1 + \alpha)n\right] E(a^\varepsilon \theta^{\alpha\varepsilon}). \qquad (6.8)$$

This same result can be achieved by imposing a uniform charge on exposure. Let the charge rate be t_d. Then the ith firm minimizes $a_i x_i^{-\alpha} + t_d x_i \theta_i$, yielding

$$x_i^* = \left[\frac{\alpha a_i}{(t_d \theta_i)}\right]^\varepsilon, \qquad (6.9)$$

which is identical to equation (6.7) if $t_d = \lambda$. Note that a conditional charge on emissions, where the charge rate for a firm with exposure factor θ_i is $\lambda \theta_i$, also yields the social cost-minimizing allocation. Thus equation (6.8) gives total social cost under the optimal exposure charge and the optimal conditional charge, and these two types of charges collapse to one alternative.

Derivation of the results for the other regulatory regimes proceeds in a similar fashion. For the emission standard the problem is to select $\bar{x}$, the standard, to minimize

$$S = \sum_{i=1}^{n} a_i \bar{x}^{-\alpha} + \lambda \sum_{i=1}^{n} \bar{x}\theta_i. \qquad (6.10)$$

Differentiating with respect to $\bar{x}$ yields the optimal emission standard:

$$\bar{x}^* = \left(\frac{\alpha \sum_{i=1}^{n} a_i}{\lambda \sum_{i=1}^{n} \theta_i}\right)^\varepsilon = \left[\frac{\alpha E(a)}{\lambda E(\theta)}\right]^\varepsilon. \qquad (6.11)$$

Substitution yields

$$S_0^* = \left[\left(\frac{\lambda}{\alpha}\right)^{\alpha\varepsilon} (1 + \alpha)n\right] [E(a)]^\varepsilon [E(\theta)]^{\alpha\varepsilon}. \qquad (6.12)$$

Note that equation (6.12) is identical to equation (6.8), the expression for social cost under the exposure/conditional charge, except that in equation (6.12) the expectations of a and θ are taken before they are raised to the

Table 6.1 Comparative social costs for exposure model

Regulatory alternative	Social cost
Emission standard	$[E(a)]^{\varepsilon}[E(\theta)]^{\alpha\varepsilon}$
Conditional standard	$E_{\theta}\{[E_{a}(a|\theta)]^{\varepsilon}\theta^{\alpha\varepsilon}\}$
Exposure standard	$[E(a\theta^{\alpha})]^{\varepsilon}$
Emission charge	$[E(a^{\varepsilon}\theta)]^{\alpha\varepsilon}[E(a^{\varepsilon})]^{\varepsilon}$
Exposure/conditional charge	$E(a^{\varepsilon}\theta^{\alpha\varepsilon})$

Note: The multiplicative constant, $(\lambda/\alpha)^{\alpha\varepsilon}(1+\alpha)n$, is omitted from each expression. $E_{a}(a|\theta)$ means expectation of a given θ.

powers $\varepsilon = 1/(\alpha+1)$ and $\alpha\varepsilon = \alpha/(\alpha+1)$, respectively. As both of these exponents are less than unity, social cost is higher under the emission standard than under the exposure/conditional charge, so long as there is any variance in the values of a and θ across sources.

Derivations for the remaining regulatory alternatives are omitted for the sake of brevity. The expression for social cost under each regime includes the multiplicative term $\left[\left(\frac{\lambda}{\alpha}\right)^{\alpha\varepsilon}(1+\alpha)n\right]$, so that the term may be omitted in comparing relative social costs, as shown in table 6.1. In each case both total damages and control costs are proportional to total social cost, as in chapter 2. That is, if with these cost and damage functions the level of pollution damage under one regime is 25 percent higher than under another, its cost of control also will be 25 percent higher.

Results with the Log-Normal Distribution

The expected social costs in table 6.1 depend on the joint distribution of a and θ across firms. The choice of such a distribution is limited severely by the need to obtain closed-form expressions for the expectation of the product of a and θ, each raised to a fractional power. Such expressions can be derived if we assume that the natural logarithms of a and θ have a bivariate normal distribution. As in the application of the univariate log-normal distribution in chapter 2, these expressions are obtained through the use of the moment-generating function.

Derivations
The moment-generating function, $M(v_1, v_2)$, for the joint distribution of two variables, z_1 and z_2, is defined as

$$M(v_1, v_2) = E[\exp(v_1 z_1 + v_2 z_2)], \tag{6.13}$$

where v_1 and v_2 are any real numbers. If we define $z_i = \ln Z_i (i = 1, 2)$, then it follows that

$$M(v_1, v_2) = E(Z_1^{v_1} Z_2^{v_2}),$$ (6.14)

where the right-hand side of equation (6.14) is of the same form as the expectations in table 6.1. If z_1 and z_2 have a bivariate normal distribution,

$$M(v_1, v_2) = \exp\left(v_1\mu_1 + v_2\mu_2 + \frac{v_1^2\sigma_1^2 + 2\rho v_1 v_2 \sigma_1 \sigma_2 + v_2^2\sigma_2^2}{2}\right),$$ (6.15)

where μ_i = the mean of z_i, σ_i^2 = the variance of z_i, and ρ = the correlation between the two.

Using equation (6.15), if we assume that $\ln a$ and $\ln \theta$ have a bivariate normal distribution, the expectation shown in table 6.1 for the exposure/conditional charge may be written as

$$E(a^\varepsilon \theta^{\alpha\varepsilon}) = \exp\left[\varepsilon\mu_a + \alpha\varepsilon\mu_\theta + \frac{\varepsilon^2(\sigma_a^2 + 2\rho\alpha\sigma_a\sigma_\theta + \alpha^2\sigma_\theta^2)}{2}\right],$$ (6.16)

where μ_a and μ_θ are the means and σ_a^2 and σ_θ^2 are the variances of $\ln a$ and $\ln \theta$, respectively, and ρ is the correlation between the two. Similarly, the relative social cost for the emission standard may be written

$$[E(a)]^\varepsilon [E(\theta)]^{\alpha\varepsilon} = \exp\left[\varepsilon\mu_a + \alpha\varepsilon\mu_\theta + \frac{\varepsilon(\sigma_a^2 + \alpha\sigma_\theta^2)}{2}\right].$$ (6.17)

Dividing equation (6.16) by equation (6.17) yields social cost under the exposure/conditional charge as a fraction of that under the emission standard:

$$\frac{S^*}{S_0^*} = \exp\left[-0.5\alpha\varepsilon^2(\sigma_a^2 - 2\rho\sigma_a\sigma_\theta + \sigma_\theta^2)\right]$$

$$= \exp\left[-0.5\alpha\varepsilon^2\sigma^2(1 - 2\rho\gamma^{0.5}(1 - \gamma)^{0.5})\right],$$ (6.18)

where $\gamma\sigma^2 \equiv \sigma_a^2$ and $(1 - \gamma)\sigma^2 \equiv \sigma_\theta^2$. That is, γ is a measure of the variance of $\ln a$ relative to the variance of $\ln \theta$, and σ^2 is the sum of those variances. The larger γ, the greater is the variance in marginal cost relative to the variance in marginal damage (exposure factor).

Table 6.2 reports the relative efficiencies of the five regulatory alternatives. Each entry, when multiplied by $-0.5\alpha\varepsilon^2\sigma^2$ and then exponentiated, gives the ratio of the social cost of that alternative to the social cost of the emission standard. Thus the larger the value of the expression, the greater is the efficiency of that alternative and the smaller its relative social cost.

Table 6.2 Relative efficiencies given log-normal distribution

Regulatory alternative	Relative efficiency
Emission standard	0
Conditional standard	$1 - \gamma(1 - \rho^2) - 2\rho\gamma^{0.5}(1 - \gamma)^{0.5}$
Exposure standard	$(1 - \alpha^2)(1 - \gamma) - 2\rho(1 + \alpha)\gamma^{0.5}(1 - \gamma)^{0.5}$
Emission charge	$\gamma - 2\rho\gamma^{0.5}(1 - \gamma)^{0.5}$
Exposure/conditional charge	$1 - 2\rho\gamma^{0.5}(1 - \gamma)^{0.5}$

Note: Let the jth entry be denoted A_j, where $A_0 =$ the entry for the emission standard. Then the ratio of social costs is $S_j^*/S_0^* = \exp[-0.5\alpha\sigma^2\varepsilon^2(A_j)]$.

Ordinal Comparisons

Using the entries in table 6.2, we can specify the ranking of regulatory alternatives under different conditions. Except in the case of the exposure standard, the only relevant parameters are ρ (the correlation coefficient), which can vary over the range $-1 \le \rho \le +1$, and γ (measures the variance in costs relative to the variance in exposure), which can vary over the range $0 \le \gamma \le 1$. Ordinal comparisons involving the exposure standard depend also on α (the elasticity of control costs with respect to emissions), which can take on any positive value.

Given five alternatives, there are ten possible paired comparisons and 120 possible different complete rankings even if ties are ignored. Many of the possibilities, however, can be eliminated at the outset. As shown in the general analysis earlier, the exposure/conditional charge always achieves the least-cost outcome, hence none of the other alternatives ever will be preferred to it. For example, the emission charge always leads to higher social cost than the exposure/conditional charge unless $\gamma = 1$, in which case the two types of charges are equivalent because the exposure factor is constant across plants. Conversely, the conditional standard matches the performance of the exposure/conditional charge only if $\gamma = 0$, in which case there is no variation in control costs, or if $\rho = \pm 1$, in which case all firms sharing the same exposure factor (θ) have the same costs.

It is also clear that neither the emission standard nor the exposure standard ever will be more efficient than the conditional standard. Intuitively, the conditional standard never can do worse than either of the other standards. The reason is the regulator always has the option of setting the same emission limit regardless of the exposure factor, in which case it is equivalent to the emission standard, or of adjusting the conditional emission limits to make exposures constant across sources, in which case it is equivalent to the exposure standard.

The other ordinal comparisons are less obvious and depend on the values of ρ, γ, and, in some cases, α. Four pairs of alternatives remain to be compared: the emission standard vs. the exposure standard, and the emission charge vs. each of the three types of standards. The four panels of figure 6.1 make these comparisons graphically, showing the values of ρ and γ for which each alternative is preferred to another. In the first two panels the comparisons also depend on α; the solid lines show the boundaries for $\alpha = 1$, the dotted lines for $\alpha = 0.5$.

If a charge is employed as the regulatory instrument, exposure always will be a better target than emissions. If, however, the instrument is a uniform standard, the choice between emissions and exposure as the basis for regulation is less clear-cut, depending on the relationship across sources between costs and exposure factors. We can gain some insight into the nature of this dependence by considering the optimal allocation and then comparing it to the results obtained under each of the two types of standards.

The optimal level of emissions at a particular source depends on two factors: (1) the cost of control, with higher costs leading to higher emissions, and (2) the exposure factor, with a higher value leading to lower emissions. If costs and exposure factors are positively correlated (i.e., if on average emissions from high-cost firms cause higher exposures per unit then those from low-cost firms), these two effects will act in opposite directions, tending to result in relatively uniform emission levels and widely varying exposure levels. In contrast, if the correlation is negative (i.e., if on average low-cost firms have high exposure factors), the two effects operate in concert, tending to increase the dispersion of emission levels and to decrease the variability in exposure levels. This line of argument suggests that the attractiveness of the emission standard relative to an exposure standard should be a positive function of the correlation coefficient, ρ.

The results in figure 6.1a are consistent with these intuitions. For $\alpha = 1$, the relationship is particularly simple; if $\rho > 0$, the emission standard is preferred; if $\rho < 0$, the exposure standard is more efficient. Decreasing the elasticity of control costs to $\alpha = 0.5$ increases the range over which the exposure standard is preferred, extending all the way to $\rho = 1$ if γ (the relative variance in costs) is small enough.

Figure 6.1b compares the exposure standard to the emission charge. Here one expects all three parameters to be important. As in the comparison with the emission standard, the exposure standard becomes more

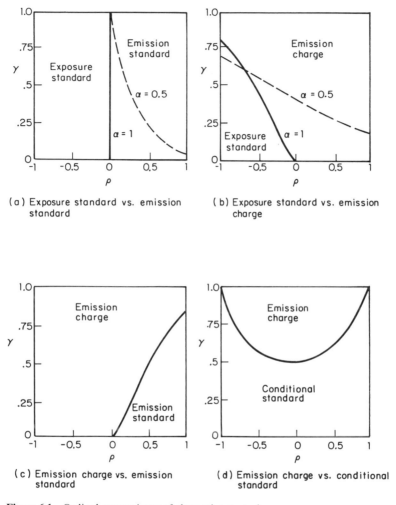

(a) Exposure standard vs. emission standard

(b) Exposure standard vs. emission charge

(c) Emission charge vs. emission standard

(d) Emission charge vs. conditional standard

Figure 6.1 Ordinal comparisons of alternative strategies

attractive as the correlation coefficient becomes more negative. Its efficiency relative to the emission charge also increases as the relative variance in costs (γ) decreases; if the variance in costs is relatively low, the ability of the emission charge to allocate higher control efforts to low-cost firms becomes less important. The elasticity of control costs (α) plays a similar role; the less elastic control costs, the less critical it is to equalize marginal costs. As shown in figure 6.1b, for $\alpha = 1$ the emission charge is superior to the exposure standard unless ρ is negative and large or γ is small. If $\alpha = 0.5$, however, the region over which the exposure standard is more efficient expands considerably; if γ is small, the exposure standard is preferred regardless of the value of ρ.

Intuitively, one might expect the emission charge to be superior to the emission standard in all cases, given the charge's ability to distribute the cleanup burden among firms on the basis of control costs. If costs and exposure factors are distributed independently across firms (i.e., if high-cost firms have the same average exposure factor as low-cost firms), then clearly it will be more efficient to have low-cost firms control more, as they will under an emission charge, than to compel all firms to meet the same emission standard. If, however, costs and exposure factors are positively correlated, then it is no longer clear that the emission charge will be superior. As the correlation between costs and exposure factors increases, at some point it may become more efficient to impose a uniform emissions level. The smaller the variance in costs relative to the variance in exposure factors, the more likely that switch is to occur. Figure 6.1c illustrates the results for the specific model used here; for $\rho \leq 0$, the emission charge is superior regardless of the value of γ. If $\rho > 0$, however, the emission standard is preferred, provided γ is "small enough", that is, provided the variance in costs is relatively small. The boundary is invariant with respect to α.

The final comparison is between the emission charge and the conditional standard. Because the conditional standard is always at least as efficient as the other two standards, it must be preferred to the emission charge in any case where the charge is inferior to either the emission standard or the exposure standard. In addition we know that the conditional standard should be most attractive when the variance in costs is low relative to the variance in exposure factors, because it allows the regulator to take account of interfirm differences in exposure factors, but not costs.

As shown in Figure 6.1d, the conditional standard is more efficient than the emission charge for the particular functional forms used if the variance in exposure factors is greater than the variance in control costs ($\gamma < 0.5$). If the variance in control costs is relatively large, however, the emission charge is superior, unless the correlation between costs and exposure factors (ρ) is quite strong, either positively or negatively. The explanation for this effect is straightforward; as the absolute value of ρ increases, the variance in control costs for any given value of θ decreases, making the conditional standard more attractive.

Magnitude of Cost Differences
The ordinal comparisons tell us which regulatory alternatives will be preferred under various conditions but not how large the differences in social costs might be. The magnitudes of these differences are important for several reasons. First, the costs of setting, monitoring, and enforcing the regulation have not been included in the formal model but are likely to differ among the alternative schemes. It would be helpful to have a rough idea of whether, for example, the additional cost and complexity of establishing a set of conditional standards rather than a single uniform standard are likely to be justified by significant savings in control costs and damages. Second, in a world of scarce political resources we want to know which battles are most important to fight. Economists have long argued for a switch from emission standards to emission charges or permits. As we have seen, in some cases such a shift may not improve efficiency, whereas modifying the standards approach to take account of differences in exposure factors (the conditional standard) may yield greater benefits. Similarly, we know that an exposure/conditional charge is more efficient than one based on emissions, but unless the gains are large, it may not be worth the political problems to attempt to place differential charges on emissions.

The kinds of issues raised here do not have simple, universal answers. As we have seen, even the ordinal comparisons depend on a variety of conditions that vary from case to case. The magnitudes of the differences are likely to be even more dependent on specific circumstances. In chapter 9 such estimates are made for a very specific problem, benzene emissions from maleic anhydride plants. At this point, however, we can use the model to make some crude estimates that can be compared with the results in earlier chapters.

The emission standard serves as the base case against which the alternatives are compared. As in chapter 2, where we compared standards and charges under the assumption that the emission-damage link was invariant

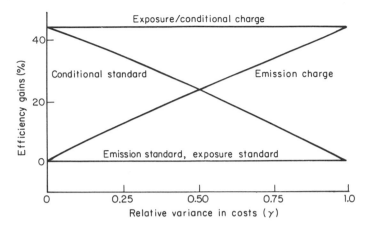

Figure 6.2   Magnitudes of efficiency gains ($\alpha = 1$)

across firms, we shall not compare the control costs of achieving a fixed level of damages or emissions but rather the net social costs of the alternatives when each is optimized. Also as before, these differences in social cost are expressed as a fraction of the cost of control in the base case, in this instance the emission standard.

Under each alternative control costs comprise the fraction $\varepsilon \equiv 1/(\alpha + 1)$ of total social cost. Thus the relative change in social cost of shifting from the emission standard to the jth alternative is given by

$$\Delta S_j^* = \frac{S_0^* - S_j^*}{\varepsilon S_0^*} = (1 + \alpha)\left(1 - \frac{S_j^*}{S_0^*}\right), \tag{6.19}$$

where $S_0^* =$ social cost under the optimal emission standard and $S_j^* =$ social cost under the alternative strategy when it has been optimized. Expressions for S_j^*/S_0^* may be derived using the entries in table 6.2. Thus, for example, shifting from the emission standard to the exposure/conditional charge yields the following relative savings in social cost:

$$\Delta S^* = (1 + \alpha)\{1 - \exp[-0.5\alpha\varepsilon^2\sigma^2(1 - 2\rho\gamma^{0.5}(1 - \gamma)^{0.5})]\}. \tag{6.20}$$

Savings for the other alternatives are given by similar expressions. In each case the value of the expression depends on four parameters: γ, α, ρ, and σ^2.

Figure 6.2 plots the savings as functions of γ, holding fixed $\rho = 0$, $\sigma^2 = 2$, and $\alpha = 1$. The net benefit of shifting from the emission standard to the exposure/conditional charge is 44 percent of the cost of control

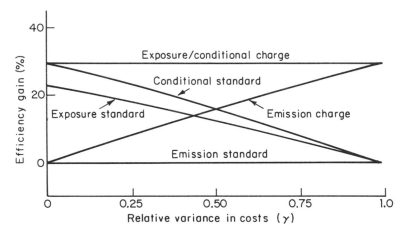

Figure 6.3   Magnitudes of efficiency gains ($\alpha = 0.5$)

under the former, regardless of the value of γ; the exposure/conditional charge is equally adept at dealing with variations in control costs and exposure factors. Note that this is the same savings for the charge found in chapter 2 (for $\sigma^2 = 2$ and $\alpha = 1$), where we assumed that marginal damages per unit of emissions were the same at all firms.

Figure 6.2 also shows that the exposure standard, for $\alpha = 1$ and $\rho = 0$, is no more efficient that the emission standard. The respective savings under the conditional standard and the emission charge depend on the relative variance. As expected, if the variance in costs is small compared to the variance in exposure factors, the emission charge does little better than the emission standard, whereas the conditional standard does almost as well as the exposure/conditional charge. If the variance in costs is high relative to the variance in exposure factors, the relationships are reversed. If the relative variances are approximately equal, the emission charge and the conditional standard perform about the same; at $\gamma = 0.5$, both yield a savings of 24 percent, as compared to 44 percent under the exposure/conditional charge.

Figure 6.3, which plots the results for $\alpha = 0.5$, shows that the exposure standard does almost as well as the conditional standard when marginal costs rise less rapidly as emissions are controlled. For the other alternatives to the emission standard, lowering α compresses the relative gains but preserves the qualitative effects of changes in γ. Note that the savings for the exposure/conditional charge, 30 percent, is the same as the savings for the charge in chapter 2 with comparable parameter values.

Additional Considerations

The results of the model suggest that targeting a charge on emissions rather than exposure may entail a substantial loss in efficiency, particularly if the variation in exposure factors is relatively large. A conditional standard or an exposure standard (when marginal control costs do not rise too rapidly as emissions approach zero) may be more efficient than a simple uniform emission charge.

No simple model can capture all of the relevant features of the problem. Thus we must be cautious in interpreting the results of the analysis. We can, however, identify some of the features not addressed by the formal model and explore, at least in a qualitative way, their implications for the choice among regulatory alternatives. Most of the issues discussed in this section have been addressed in earlier chapters, though in somewhat different contexts. The overall impact of these additional considerations is to increase the relative attractiveness of the exposure/conditional charge.

Information Requirements and Uncertainty

The results reported previously assume that all of the relevant parameter values are known with certainty. Often, however, as discussed in chapter 4, the confidence limits around the estimate of marginal damages may encompass an order of magnitude or more. The cost estimates, though generally more reliable, are rarely very precise. Thus we need to consider how errors in estimating either costs or damages may affect each regulatory alternative.

If the shadow price on damage (λ) is misestimated by the factor k_λ, it can be shown that the social cost under each regulatory alternative will rise by the factor $(k_\lambda^{\alpha\varepsilon} + \alpha k_\lambda^{-\varepsilon})\varepsilon$. This is the same factor found in chapter 4 (see equation 4.9) for the simple model that assumes constant marginal damage across sources. Thus misestimating marginal damage has the same proportional effect on each alternative, but it *increases* the absolute differences in efficiency. In particular, it widens the gap between the exposure/conditional charge and the other strategies.

If the cost coefficients are misestimated by a factor of k_a, the two types of charges still achieve the results found earlier under the assumption of perfect information. This result should not be surprising; when marginal damage does not vary with the emission level, the optimal charge rate does not depend on costs, as shown in chapter 4. The three types of standards, however, are sensitive to the accuracy of the cost estimates. More specifically, if costs are misestimated by the factor k_a, the social cost under each of the standards rises by a factor of $(k_a^{-\alpha\varepsilon} + \alpha k_a^{\varepsilon})\varepsilon$, the same factor found in chapter 4 (see equation 4.14). Thus uncertainty about costs widens the absolute differences among the three types of standards and decreases their efficiency as compared to the two types of charges.

Monitoring Problems

In theory the exposure standard and the exposure charge could be enforced by monitoring exposures on a source-by-source basis. In most cases, however, direct measurements of exposure from a particular source are not feasible, so enforcement must be based on emissions or some surrogate. The problem is to combine measured emissions with estimated exposure factors to generate source-specific estimates of total exposure levels for enforcement purposes. Once this is done, all of the alternatives entail the same type of ongoing monitoring of emissions, preferably on a continuous, direct measurement basis, but alternatively using periodic samples or combining test data with records of production levels, as discussed in the previous chapter.

All of the regulatory alternatives require estimates of exposure factors. For the charge or standard targeted on emissions, however, the exposure factors need not be source specific; averages generally will be sufficient. For the conditional standard, source-specific data are required, but the estimates need not be very precise if the number of exposure categories is kept small, with each category encompassing a sizable range of exposure factors. Limiting the number of categories for the conditional standard also is desirable from the standpoint of limiting the number of separate cost estimates that need to be made. There is, however, a trade-off; the smaller the number of categories, the less precise the effect of the conditional standard.[4]

Reasonably accurate estimates of source-specific exposure factors are likely to be most important for the exposure standard and the exposure/conditional charge. Chapter 9 presents examples of such estimates prepared by EPA for its analysis of the health effects of benzene emissions from maleic anhydride plants. Those estimates employ only source-specific population data; the same "average" meteorological parameters and plant-operating characteristics are used for the dispersion modeling around all plants. Where possible, however, it would be desirable to employ meteorological data that reflect more accurately the weather conditions in a given plant's region. If such modeling were done, it would be a relatively simple matter to design a computer program that would combine the dispersion model results with population data to generate an exposure factor for any given site. Note also that, once developed, such a program could be used in designing regulations for many different substances. If those estimates were updated periodically, they would provide an automatic process for adjusting an exposure charge in response to changes in population patterns.

Prices and Revenues

All of the alternatives will tend to drive up the prices of the final goods produced, as all will raise production costs. Only the exposure/conditional charge will provide the optimal price signals, however, because only in that case will each firm pay a charge equal to the residual damage it imposes. Under the uniform emission charge the total charge paid by any individual firm will be too low or too high, depending, respectively, on whether its exposure factor is higher or lower than average. Under all of the standards, of course, firms pay nothing for the residual damages imposed, so the price to consumers generally will be too low. As discussed in chapter 3, the importance of these differential impacts on costs will depend on a variety of factors, including the elasticities of supply and demand, market structure, and the value of residual damage relative to the cost of the product.

The standards also suffer from the disadvantage that they generate no revenues and hence yield no tax-displacement benefits. Both the emission charge and the exposure/conditional charge, in contrast, may yield substantial revenues, contributing further to their efficiency advantages. As shown in chapter 3, the benefits of using charge revenues to displace deadweight losses from preexisting taxes may be quite substantial, perhaps even on a par with the benefits of an improved allocation of control efforts. Revenues under the emission charge generally will be somewhat higher than under the exposure charge, primarily because the former leaves a higher level of residual damages. Thus the general effect of taking account of tax-displacement benefits is to increase the attractiveness of the charges relative to the standards and to narrow slightly the efficiency gap between the two types of charges.

Location Incentives

In the formal model we assume that each source's site is fixed. That assumption is quite reasonable for the short run but may not be for the long run, when firms have the option of relocating existing plants and new or existing firms must decide where to locate new plants. Plant location may become an economical alternative to emission control for reducing

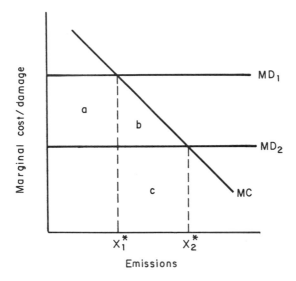

Figure 6.4   Relocation incentives under alternative instruments

exposure and hence damages. As shown in chapter 9, for example, moving the maleic anhydride plant in the highest-damage location to the site of the plant with the lowest exposure factor would cut that plant's total exposure level by almost 98 percent, even if emissions were held constant.

The exposure/conditional charge provides the appropriate incentives for firms to locate at low-damage sites. In the example of a firm choosing between the two sites, it could reduce its charge payments 98 percent by choosing the low-damage site. Under a uniform emission charge, however, the firm has no incentive to choose the low-damage site because its emission charge payments would not be affected. The emission standard also fails to provide any incentive to locate in low-damage areas.

The conditional and exposure standards provide incentives for low-damage siting, but not necessarily at the appropriate levels. Figure 6.4 illustrates the problem for a firm choosing between two alternative sites. Its marginal cost of control is given by the curve MC. At its current site, location 1, marginal damage is MD_1 and the optimal level of emissions is x_1. At the alternative, location 2, marginal damage is lower, MD_2, and the optimal level of emissions is x_2. Relocating reduces control costs by the area $b + c$ and total damages by $a - c$. Thus switching sites will yield positive net benefits if the cost of moving is less than $a + b$. Under an exposure/conditional charge the external damages are internalized, so the firm faces precisely the right incentive; moving reduces control costs by $b + c$ and cuts charge payments by $a - c$. With the standards, however, even if they are set at precisely the right level at each site, the incentive for relocation is not likely to be optimal; the gain from moving to the new site is the reduction in control costs, $b + c$, less moving costs. Thus, unless $a = c$, the incentive will not be optimal. If the demand for emissions (the MC curve) is inelastic, $a > c$ and the standard provides an inadequate incentive to relocate. If demand is elastic, $a < c$ and the standard provides an excessive incentive to move. Only if demand is unit elastic will $a = c$ and the incentive be just right.

Equity Considerations
The regulatory strategies analyzed in this chapter vary not only in their efficiency but also in how they distribute costs and benefits across individuals and firms. Switching from an emission standard to an exposure charge, for example, will raise the profits of some firms but depress those of others. Similarly, it will decrease the risks faced by some individuals but raise those faced by others. Although a detailed distributional analysis is impossible at this point, given the general nature of the analysis, we can explore some of the equity issues involved.

When initially presented with the idea of exposure-based charges or standards, may individuals raise a variety of equity-based objections. On closer examination, however, most of these objections prove groundless. The most common objection arises when the exposure/conditional charge is presented in terms of imposing a higher charge on emissions from sources in densely populated areas. The concern is that such an approach discriminates against people who live in lightly populated rural areas, that it places a lower value on protecting their health. In fact, however, a uniform exposure charge—which, as shown earlier, is analytically equivalent to varying an emission charge in proportion to exposure factors—is based on the notion that the social benefit derived from protecting an individual is the same regardless of where that individual lives. In contrast, a uniform emission charge (or a uniform standard) implicitly assigns radically different weights to protecting different individuals. If one plant has an exposure factor 100 times that of another, for example, a uniform emission charge implies that the value of protecting an individual living near the first plant is 1 percent the value of protecting an individual living near the second. Thus, if the criterion for equity is that lives be valued equally, emission-based strategies are distinctly inferior to those targeted on exposure.

Another possible measure of equity is the extent to which risks are distributed reasonably equally and are not concentrated among a few individuals. This criterion is impossible to apply rigorously, as there always will be variations in exposure and individual susceptibility, and hence in risk, regardless of the regulatory approach taken. Moreover it makes little sense to apply that criterion on a source-by-source or substance-by-substance basis. That is, the distribution of risks posed by benzene emitted from, say, maleic anhydride plants is of less importance than the distribution of risks from hazardous environmental pollutants generally.

In many cases exposure-based approaches actually will lead to a more even distribution of risk than will uniform emission standards or charges. If a uniform emission standard is imposed, the emissions per unit of capacity will be constant, but the ambient concentration in an area will depend on the sizes and numbers of sources. As many emission sources, particularly automobiles, tend to be located where people are, those who live in densely populated areas will face higher exposures because of the greater number of sources in their vicinity. In contrast, an exposure charge or standard will tend to lead to tighter controls on sources in densely populated areas, hence counteracting, at least partially, the effect of a greater number of sources. Thus, rather than leading to greater variability in risk levels, exposure charges may reduce the variance in exposure levels relative to that achieved under uniform emission charges or standards.

Still another potential equity-based objection to the exposure-charge concept, or to its standard-based counterparts, is that they will have differential impacts on otherwise identical firms, depending on their locations. Coupled with that objection may be the fear that high-exposure urban areas will be unfairly disadvantaged in attracting and retaining industry, given that a firm could lower its charge rate or meet a looser standard by choosing a site in a less densely populated area.

The assertion that an exposure-based strategy will have differential effects on sources located in different areas, and that these differentials will affect at least some locational decisions, is undeniable. Indeed, as argued earlier, these effects increase its efficiency. That these effects will be unfair or in some way undesirable is far from clear. No form of regulation affects all firms equally. Whether the regulation employs charges, permits, or standards, and whether it is based on emissions or exposure, some firms bear higher costs than others. Nor is it clear that it is inequitable for a firm in a high-damage area to pay a higher charge for its emissions than a firm in a low-damage area does, any more than it is unfair that a firm located in New York City has to pay higher rents for its facilities than it would if it chose to locate in a less densely populated area. In both cases the prices reflect the opportunity costs.[5]

Implications for Marketable Permits

Under conditions of certainty suitably modified marketable permit schemes can provide the same results as the exposure/conditional charge. One approach is to issue different sets of emission permits based on exposure factors and to restrict trading to plants with the same exposure factors. Another approach is to define the permits in terms of exposure rather than emissions; a source with an exposure factor of 0.2 units of exposure per kilogram of emissions, for example, could emit 5,000 kg for every 1,000 permits, whereas a source with an exposure factor of 0.5 could emit only 2,000 kg with the same number of permits. If expected marginal damages are fairly constant, the second approach will be preferable because it does not require separate cost estimates for plants with different exposure factors. Neither system of permits will be as robust as an exposure charge, however, in the face of uncertainty about costs.

Summary

If it is to be fully efficient, a regulatory strategy must deal with variations in both the marginal benefits and the marginal costs of controlling emissions. A uniform emission standard does neither. A uniform emission charge, though it deals with differences among firms in the costs of controlling emissions, fails to reflect differences in marginal damages. In contrast, both exposure standards and emission standards conditional on exposure factors take account of differences in marginal damages, but neither copes with variations in cost. Only a charge targeted on exposure (or some other measure that is a close proxy for damages) will achieve full efficiency.

The efficiency advantage of the exposure charge over the other alternatives may be quite large. The model employed in this chapter suggests that under many conditions the gain in shifting the target of a charge from emissions to exposure is larger than that of shifting from an emission standard to an emission charge. Uncertainty, impacts on the prices of final goods, tax-displacement benefits, and location incentives, although not included in the formal model, all strengthen the case for the exposure charge. Moreover these efficiency gains do not appear to be bought at the price of inequity; under the exposure charge firms bear the costs of the external damages they impose, and the implicit shadow price on protecting the individual is uniform.

Chapter 7

Multiple Control Variables: Specification vs. Performance Targets

Regulators unable to monitor source-specific damages directly frequently are able to measure key variables that determine damage levels. In some cases these individual measurements can be combined to yield source-specific damage estimates that may serve as the target of regulation. In other cases, however, it is difficult to assemble the data necessary to yield a comprehensive target, and several separate targets must be employed.

Consider the problem of automobile emissions, which cannot be monitored on a continuous, source-specific basis. More than a decade ago economists at the Rand Corporation suggested a rather ingenious way around this problem (Kneese and Schultze 1975, 101–102). They proposed that each vehicle be tested periodically and given a seal indicating its emission rate. With each purchase of gasoline, the owner would be assessed a charge based on the vehicle's emission rate and the number of gallons purchased. In essence, the plan would combine two measures, the emission rate and gallons of gas used, to construct a target that would be a very close proxy for actual emissions. Benzene emissions from service stations offer similar opportunities for constructing a comprehensive target; as discussed briefly in chapter 5, station-specific emissions can be estimated by measuring the benzene content of the gasoline and the hydrocarbon emission rate per gallon for each station.

It may be difficult, however, to match the individual measurements on a source-by-source basis. This problem is particularly severe for the Rand proposal; the regulator would have difficulty monitoring compliance by service stations, making sure that they were in fact imposing the appropriate charge rates on cars with different emission ratings. Owners of cars with high emission rates, for example, might bribe station attendants to assess at a lower rate, and the regulatory agency would have a hard time proving that such cheating had occurred. Monitoring might be less difficult in the case of benzene emissions from service stations, though even there the administrative problems of keeping track of which service stations received gasolines with varying benzene contents could be significant.

In both cases it would be easier administratively to have separate targets. For example, individuals could be assessed a charge based on the emission ratings of their cars and a uniform charge could be levied on gasoline to provide an incentive for people to drive less. Similarly, a charge could be imposed on refiners based on the benzene content of gasoline, and another charge could be imposed on service stations based on evaporative controls and number of gallons pumped. Analogous options would be available with standards. Alternatively, a charge might be used with one target, a standard with another. An emission standard for cars, for example, might be coupled with a charge (tax) on gasoline.

The administrative ease of separate targets is gained at the cost of some inefficiency, as discussed briefly in chapter 5. Performance measures, though sometimes more difficult to monitor and enforce, give firms more flexibility in reducing damages. With the aid of models similar to those used in earlier chapters, we can gain some insight into the efficiency advantages of performance measures and into how the choice of instrument affects the importance of using performance targets.

The General Model

Suppose we have n different sources, each emitting some hazardous substance. The damage caused by the ith firm, d_i, is a function of two variables, x_i and y_i:

$$d_i = D(x_i, y_i), \tag{7.1}$$

where $D(\cdot)$ is the same across sources, but x_i and y_i are both under the control of the individual source. The marginal damage due to each variable is positive ($D_x^i > 0$, $D_y^i > 0$) and, to keep matters simple, constant ($D_{xx}^i = 0$, $D_{yy}^i = 0$). Increasing the value of one variable, however, may increase the marginal damage due to the other ($D_{xy}^i \geq 0$). The greater the emission rate for a given car, for example, the greater the damage caused by another mile of driving. Let the costs be given by $C^i(x_i, y_i)$, where $C^i(\cdot)$ varies across firms. The marginal costs of controlling the variables are positive ($-C_x^i > 0$, $-C_y^i > 0$), increasing ($C_{xx}^i > 0$, $C_{yy}^i > 0$), and independent ($C_{xy}^i = 0$).

If the costs and damages are additive across firms, social cost is given by

$$S = \sum_{i=1}^{n} C^i(x_i, y_i) + \sum_{i=1}^{n} D(x_i, y_i). \tag{7.2}$$

The first-order conditions for minimum social cost are

$$-C_x^i = D_x(x_i, y_i), \tag{7.3a}$$

and

$$-C_y^i = D_y(x_i, y_i), \quad i = 1, \ldots, n. \tag{7.3b}$$

Equations (7.3a) and (7.3b) simply state that each source should control each variable to the point where the marginal cost of control is equal to marginal damage. This optimum can be achieved by levying a uniform performance charge directly on damages. Specification charges levied on x and y separately generally will not be fully efficient. If separate charges are levied on x and y at the rates t_x and t_y, respectively, cost minimization by firms will lead to the following conditions:

$$-C_x^i = t_x \tag{7.4a}$$

and

$$-C_y^i = t_y, \quad i = 1, \ldots, n. \tag{7.4b}$$

That is, each specification charge will equalize across sources the marginal cost of controlling that variable. At the optimum, however, as we can see from equations (7.3a, b), the marginal cost of controlling one variable will vary across firms unless the other variable has the same value at all sources, or the marginal damage caused by each variable is independent of the level of the other (i.e., $D_{xy}^i = 0$). Thus uniform specification charges will be suboptimal in many instances.

Separate specification standards also will be less efficient than a unified performance standard. Suppose that each firm must meet separate standards, $x_i \leq \bar{x}$ and $y_i \leq \bar{y}$. We could achieve the same damage level with a performance standard, $\bar{d} = D(\bar{x}, \bar{y})$, which would allow each source to find the mix of control variable values that minimized its own costs.

To proceed further, we need to specify the structures of the cost and damage functions in more detail and to make some assumptions about the distributions of the parameters of those functions. The next section specifies particular cost and damage functions similar to those used earlier. The following section uses the bivariate log-normal distribution to derive more specific results.

Specific Cost and Damage Functions

Let the damages from each source be proportional to the product of the two control variables:

$$D(x_i, y_i) = \lambda x_i y_i, \quad i = 1, \ldots, n, \tag{7.5}$$

where λ is again the shadow price of the damage measure. This functional form is appropriate in many cases. In the auto emissions example x_i could represent the ith car's emission rate per gallon of gas consumed, and y_i would represent the number of gallons consumed. The product $x_i y_i$ would then be that car's emissions. Similarly, in the case of benzene emissions from service stations, x_i could represent the benzene content of the gasoline and y_i the emission factor for the station. The shadow price, λ, in both cases would represent the dollar value of reducing emissions. This assumes, of course, that the marginal damage caused by emissions does not vary across sources, as might be the case if we were dealing only with sources in a relatively small area. Alternatively, we can view this formulation as an extension of the model employed in chapter 6. In that case we might interpret x_i as the ith source's emissions and y_i as its exposure factor, where we now assume that firms can, at some cost, reduce their exposure factors by relocating.

Let the cost for the ith firm be given by

$$C^i(x_i, y_i) = a_i x_i^{-2\alpha} + b_i y_i^{-2\alpha}. \tag{7.6}$$

That is, the (absolute value of the) elasticity of control costs is constant for each variable and is equal to 2α. If a firm is allowed to pick the least-cost mix for achieving any given reduction in damages, it can be shown that the elasticity of total control cost with respect to damages is α, as in earlier chapters. The intuitive explanation is that if we reduce x_i and y_i each by 0.5 percent, total costs will rise by $2\alpha(0.5) = \alpha$ percent, and their product will fall by $(0.5 + 0.5) = 1$ percent.

The Options
Let us consider five regulatory alternatives: (1) *specification standards* that set separate standards for the two variables (e.g., gasoline can contain no more than $\bar{x}$ percent benzene, and service stations must have an emissions factor of no more than $\bar{y}$); (2) a *performance standard* that imposes a uniform minimum level of performance but allows each source to choose its own mix of control values (e.g., no service station shall emit more than $\bar{d}$ grams of benzene per gallon of gas); (3) a *combined specification charge and standard*, whereby a standard is imposed on one variable, y, but a charge is levied on the other, x (e.g., refiners must pay a charge of t_x per kilogram of benzene in gasoline, and service stations must have an emission factor of no more than $\bar{y}$); (4) *specification charges*, whereby a separate charge is levied on each variable (e.g., refiners will pay a charge of t_x per kilogram of benzene, and service stations will pay a charge of

t_y per kilogram of hydrocarbon emissions, regardless of their benzene content); and (5) a *performance charge* that imposes a uniform charge on $x_i \theta_i$ (e.g., each service station will pay a charge of t_d per kilogram of benzene emitted).

Derivations

As before, the goal is to minimize social cost:

$$S = \sum_{i=1}^{n} (a_i x_i^{-2\alpha} + b_i y_i^{-2\alpha}) + \lambda \sum_{i=1}^{n} x_i y_i, \tag{7.7}$$

where the first summation is total control costs and the second is total external damages. Differentiating with respect to each of the x_i's and y_i's, we obtain the first-order optimality conditions:

$$\frac{\partial S}{\partial x_i} = -2\alpha a_i x_i^{-2\alpha-1} + \lambda y_i = 0 \tag{7.8a}$$

and

$$\frac{\partial S}{\partial y_i} = -2\alpha b_i y_i^{-2\alpha-1} + \lambda x_i = 0, \quad i = 1, \ldots, n. \tag{7.8b}$$

Solving these equations simultaneously yields the optimal values of the control variables:

$$x_i^* = \left(\frac{2\alpha}{\lambda}\right)^{0.5\varepsilon} \left(\frac{a_i^{(2\alpha+1)}}{b_i}\right)^{0.25\varepsilon/\alpha} \tag{7.9a}$$

and

$$y_i^* = \left(\frac{2\alpha}{\lambda}\right)^{0.5\varepsilon} \left(\frac{b_i^{(2\alpha+1)}}{a_i}\right)^{0.25\varepsilon/\alpha}, \quad i = 1, \ldots, n, \tag{7.9b}$$

where $\varepsilon \equiv 1/(\alpha + 1)$, as in earlier chapters. Minimum social cost is then

$$S^* = \left[2^\varepsilon \left(\frac{\lambda}{\alpha}\right)^{\alpha\varepsilon} n(1 + \alpha)\right] E(ab)^{0.5\varepsilon}. \tag{7.10}$$

A uniform performance charge at the rate $t_d = \lambda$ achieves this result. We can interpret ε as the (absolute value of the) elasticity of demand for the right to impose external damages with respect to the performance charge.

In the interests of brevity, derivations for the other regulatory alternatives are omitted. Table 7.1 summarizes the results for all five alternatives. The parameter $\eta \equiv 1/(2\alpha + 1)$ may be interpreted as the (absolute value to the) elasticity of demand for each variable with respect to a specifica-

Table 7.1   Comparative social costs with two control variables

Regulatory alternative	Social cost
Specification standards	$[E(a)E(b)]^{0.5\varepsilon}$
Performance standard	$[E(ab)^{0.5}]^{\varepsilon}$
Specification charge-standard	$\{[E(a^{\eta})]^{1/\eta}E(b)\}^{0.5\varepsilon}$
Specification charges	$\{E(a^{\eta})E(b^{\eta})[E(ab)^{\eta}]^{2\alpha}\}^{0.5\varepsilon}$
Performance charge	$E(ab)^{0.5\varepsilon}$

Note: Multiplicative constant, $(\lambda/\alpha)^{\alpha\varepsilon}2^{\varepsilon}n(1+\alpha)$, omitted from each expression.

Table 7.2   Relative efficiencies of two control variable model given log-normal distribution

Regulatory alternative	Efficiency
Specification standards	$-1 + 2\rho\gamma^{0.5}(1-\gamma)^{0.5}$
Performance standard	0
Specification charge-standard	$-1 + 4\alpha\eta\gamma + 2\rho\gamma^{0.5}(1-\gamma)^{0.5}$
Specification charges	$1 - 2\eta + 2(1 - 4\alpha\eta^{2})\rho\gamma^{0.5}(1-\gamma)^{0.5}$
Performance charge	$\alpha\varepsilon[1 + 2\rho\gamma^{0.5}(1-\gamma)^{0.5}]$

Note: If the jth entry is denoted A_j, with A_0 = the entry for the performance standard, then $S_j^*/S_0^* = \exp(-0.125\varepsilon\sigma^2 A_j)$.

tion charge imposed on that variable. As in earlier chapters, control costs and damages under each alternative are both proportional to total social cost, with control costs making up the fraction $\varepsilon \equiv 1(1+\alpha)$ and damages comprising the remainder $\alpha\varepsilon = \alpha/(1+\alpha)$.

Results with the Log-Normal Distribution

We can derive some useful analytic results once again if we assume that the natural logs of the cost coefficients have a bivariate normal distribution across sources. Table 7.2 reports the relative efficiency of each alternative, normalized so that the social cost under the performance standard is unity. Analogous to the results in chapter 6, ρ is the correlation between $\ln a$ and $\ln b$, $\gamma\sigma^2$ is the variance of $\ln a$, and $(1-\gamma)\sigma^2$ is the variance of $\ln b$. As before, σ^2 is the sum of the variances, and γ is a measure of their relative levels. Each entry in table 7.2, when multiplied by $-0.125\varepsilon\sigma^2$ and exponentiated, yields the ratio of the social cost of that alternative to the social cost of the performance standard.

Ordinal Comparisons

The rankings of the alternatives depend on the values of three parameters: ρ, γ, and α (ε and η are both functions of α). Ten paired comparisons are possible, but in most of these the preferred alternative is invariant over the relevant ranges of parameter values ($\alpha > 0$, $-1 \leq \rho \leq 1$, and $0 \leq \gamma \leq 1$). The fact that the performance charge achieves minimum social cost eliminates four possible paired comparisons. At the other extreme the specification standards are never more efficient than any of the other alternatives, thus eliminating three more possibilities.

Three pairs remain to be compared: (1) performance standard vs. specification charges; (2) performance standard vs. combined specification charge and standard; and (3) specification charges vs. combined specification charge and standard. The three panels of figure 7.1 make these comparisons graphically for all possible values of γ and ρ, and for two values of α. As in figure 6.1 the solid line shows the boundary for $\alpha = 1$ and the dotted line the boundary for $\alpha = 0.5$.

As shown in panel (a), if $\alpha = 1$, the specification charges are more efficient than the performance standard unless the two cost coefficients are highly negatively correlated and neither of the cost coefficients has much greater variance than the other. If the cost functions are less sharply curved, the performance standard is more efficient over a broader range; for $\alpha = 0.5$, it outperforms the specification charges for all values of γ if the correlation is negative.

The intuitive explanations for these results are straightforward. As discussed in earlier chapters, standards are at less of a disadvantage when marginal costs do not change rapidly as the level of control increases; thus the smaller the value of α, the better the standard does. The more negative the correlation is between the cost coefficients, the less the variation in performance levels at the optimum; firms that have higher than usual costs for one variable will have, on average, lower than usual costs for the other variable, so that imposing a uniform performance level will not be terribly inefficient.

Much the same reasoning explains the results in figure 7.1b, which compares the performance standard and the combination of a charge on x and a standard on y. Again the performance standard does better the less sharply nonlinear the cost functions and the more negative the correlation between the two cost coefficients. The combination charge-standard does better the larger the value of γ; with high values of γ, the standard is imposed on the variable with little variation in costs, y. Given a fixed value of y across firms, the marginal damage caused by x is constant, and the charge levied on that variable is fully efficient in allocating control efforts with regard to that variable.

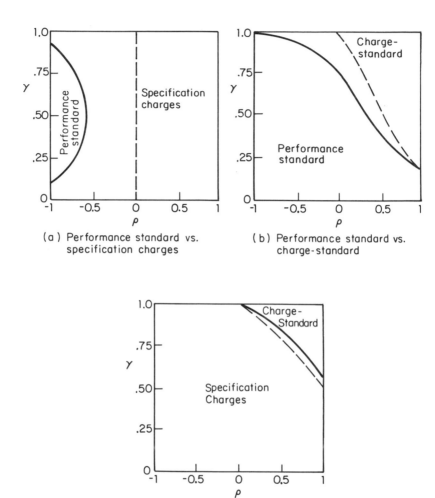

(a) Performance standard vs. specification charges

(b) Performance standard vs. charge-standard

(c) Specification charges vs. charge-standard

Figure 7.1 Ordinal comparisons with two-variable model

The same pattern shows up in panel c; the hybrid approach does relatively well, this time compared to specification charges, when γ is large and ρ is positive. The explanation is that when costs are positively correlated, the specification charges cause, on average, too little control of both variables at high-cost sources and too much at low-cost sources. Switching to a standard on y tightens controls on that variable at high-cost sources and loosens them at low-cost sources. The result is, on average, too much control of y at high-cost sources and too little at low-cost sources. If the cost of controlling y varies little across sources, however, the discrepancy will not be great, and as noted previously, once y is fixed by a standard, the charge on x allocates control of that variable efficiently. Thus in a limited number of cases efficiency can be improved by switching from pure specification charges to a hybrid approach.

Magnitudes of Cost Differences
As in chapter 6, the ordinal comparisons tell us the conditions under which one alternative will be more efficient than another but nothing about the magnitudes of the differences. It is important to have a rough idea of whether the efficiency losses entailed by employing separate specification targets rather than unified performance measures are trivial or major.

As in previous chapters, let us measure the change in total social cost of switching from a base case, here the performance standard, to another alternative as a fraction of the control costs incurred under the base case. Note also, as before, that we are comparing the social costs of the different strategies when each is optimized.

Under each alternative control costs comprise the fraction $\varepsilon \equiv 1/(\alpha + 1)$ of total social cost. Thus the relative change in social cost of shifting from the performance standard to the jth alternative is

$$\Delta S_j^* = \frac{S_0^* - S_j^*}{\varepsilon S_0^*} = (1 + \alpha)\left(1 - \frac{S_j^*}{S_0^*}\right). \tag{7.11}$$

The ratio S_j^*/S_0^* may be evaluated using the expressions in Table 7.2. Shifting from the performance standard to the performance charge, for example, yields the following relative gain in efficiency:

$$\Delta S^* = (1 + \alpha)\{1 - \exp[-0.125\alpha\varepsilon^2\sigma^2(1 + 2\rho\gamma^{0.5}(1 - \gamma)^{0.5})]\}. \tag{7.12}$$

Figure 7.2 plots the relative efficiencies of the alternatives as functions of ρ, the correlation coefficient, holding constant $\gamma = 0.5$, $\alpha = 1$, and

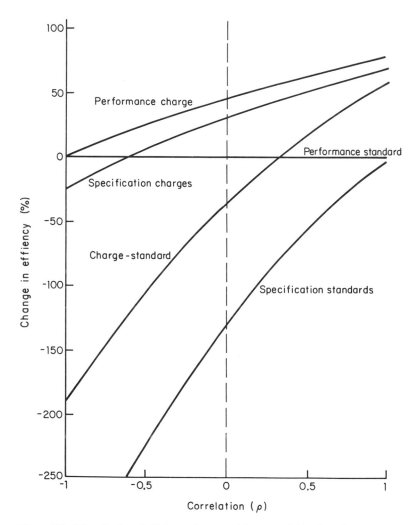

Figure 7.2 Magnitudes of efficiency changes with two variables ($\alpha = 1$)

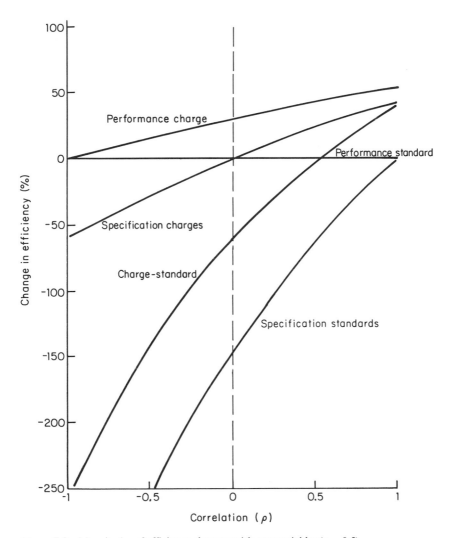

Figure 7.3   Magnitudes of efficiency changes with two variables ($\alpha = 0.5$)

$\sigma^2 = 8$. The value assumed for σ^2 may seem unreasonably high compared to earlier chapters, where $\sigma^2 = 2$. Note, however, two facts: (1) there are now two control variables rather than only one so that, if $\gamma = 0.5$, $\sigma_a^2 = \sigma_b^2 = 4$, and (2) if the two cost coefficients are not correlated, the geometric variance in the cost of achieving a given level of control will be lower than the variance associated with each individual coefficient. As shown in figure 7.2, in fact, if $\sigma^2 = 8$, and $\rho = 0$, the gain in switching from the performance standard to the performance charge is 44 percent of the control costs under the latter, exactly the same gain found for $\sigma^2 = 2$ in switching from a standard to a charge in chapter 2, and from an emission standard to an exposure charge in chapter 6, holding other parameter values constant.

The most important feature brought out by figure 7.2 is the abysmal performance of the specification standards. For $\rho = 0$, the loss in efficiency of switching from a performance standard to separate specification standards is 130 percent of the control cost under the former. The hybrid specification charge-standard does considerably better, though not as well as the performance standard unless the correlation is positive; for $\rho = 0$, the hybrid is 36 percent less efficient than the performance standard. In contrast, the specification charges do reasonably well; for $\rho = 0$, the efficiency gain is 31 percent, as opposed to 44 percent under the performance charge.

Figure 7.3 plots the results for $\alpha = 0.5$. Compared to the results in figure 7.2, all three specification-based regulations do somewhat worse; at $\rho = 0$, for example, the efficiency loss under the specification standards is 142 percent, and the specification charges yield no gain over the performance standard. For $\rho = 0$, the performance charge generates a 30 percent improvement in efficiency over the performance standard, the same gain recorded in chapters 2 and 6 for $\alpha = 0.5$ and $\sigma^2 = 2$.

Multistage Generalization

The analysis of a two-stage process is easily generalized to a process with any number of stages. Let z_{ij} be the value of the jth control variable ($j = 1, \ldots, m$) at the ith source ($i = 1, \ldots, n$), and let c_{ij} be its corresponding cost coefficient. As in the two-stage model we assume that the elasticity of the cost of controlling each variable is constant across variables and sources and that the damages from each source are proportional to the product of the control variables.

Table 7.3   Comparative social costs for multistage model

Regulatory alternative	Social cost
Specification standards	$[\prod_j E(c_j)]^{\varepsilon/m}$
Performance standard	$[E(\prod_j c_j^{1/m})]^{\varepsilon}$
Specification charges	$[\prod_j E(c_j^{\eta})]^{\varepsilon/m}[E(\prod_j c_j^{\eta})]^{\alpha\varepsilon}$
Performance charge	$E(\prod_j c_j^{\varepsilon/m})$

Note: Multiplicative constant, $(\lambda/\alpha)^{\alpha\varepsilon}m^{\varepsilon}n(1 + \alpha)$, omitted from each expression. $\eta \equiv 1/(m\alpha + 1)$.

Derivations

Total social cost is given by

$$S = \sum_{i=1}^{n}\sum_{j=1}^{m} c_{ij}z_{ij}^{-m\alpha} + \lambda\sum_{i=1}^{n}\prod_{j=1}^{m} z_{ij}. \qquad (7.13)$$

Note that the (absolute value of the) elasticity of control costs with respect to any one variable is $m\alpha$, but as in the two-variable case, the elasticity of control costs with respect to damages at each firm is α. Differentiating with respect to each z_{ij} and setting the results equal to zero yields the optimal values of the control variables:

$$z_{ij}^{*} = \frac{\left(\dfrac{m\alpha}{\lambda}\right)^{\varepsilon/m} c_{ij}^{1/(m\alpha)}}{\left[\displaystyle\prod_{k=1}^{m} c_{ik}\right]^{\varepsilon/(m^2\alpha)}}, \quad i = 1, \ldots, n; j = 1, \ldots, m. \qquad (7.14)$$

As in the earlier analyses, $\varepsilon \equiv 1/(\alpha + 1)$ may be interpreted as the (absolute value of the) elasticity of demand for damages with respect to the price charged for damages. (Note that, if $m = 2$ and we substitute $x_i = z_{i1}$, $y_i = z_{i2}$, $a_i = c_{i1}$, and $b_i = c_{i2}$, equation 7.14 reduces to equation 7.9a.) Minimum social cost is given by

$$S^* = \left[m^{\alpha}\left(\frac{\lambda}{\alpha}\right)^{\alpha\varepsilon}(1 + \alpha)n\right] E\left[\prod_{j=1}^{m} c_j^{\varepsilon/m}\right], \qquad (7.15)$$

where the expectation is taken over sources. As in the two-variable case this optimum can be achieved by levying a uniform charge on performance at the rate λ. Table 7.3 compares the social costs of four alternative strategies: specification and performance standards, and specification and performance charges. The parameter $\eta \equiv 1/(m\alpha + 1)$ represents the (absolute value of the) elasticity of demand for each variable with respect to the charge imposed on that variable.

Table 7.4   Relative efficiencies of multistage model given log-normal distribution

Regulatory alternative	
Specification standards	$1 - m + V$
Performance standard	0
Specification charges	$1 - m\eta + 2V(1 - m^2\alpha\eta^2)$
Performance charge	$\alpha\varepsilon(1 + 2V)$

Note: If the jth entry is denoted A_j, where A_0 = entry for performance standard, then $S_j^*/S_0^* = \exp[-0.5\varepsilon/m^2(A_j)]$. $\eta \equiv 1/(m\alpha + 1)$.

Log-Normal Results

If we assume that the natural logs of the cost coefficients have a multivariate normal distribution across sources, we can derive some analytic results using, as before, the moment-generating function. Table 7.4 presents the relative efficiency of each alternative, normalized again to the performance standard. In these expressions, $\sigma^2 \equiv \sum_{j=1}^{m} \sigma_j^2$ is the sum of the variances, as before, and

$$V \equiv \sum_{j=1}^{m-1} \sum_{k=j+1}^{m} \rho_{jk}(\gamma_j\gamma_k)^{0.5},$$

where ρ_{jk} is the correlation between the jth and kth cost coefficients and γ_j is the jth coefficient's share of the sum of the variances ($\sum_{j=1}^{m}\gamma_j = 1$). (Thus $\sigma^2 V$ is the sum of the covariances.)

Once again the difference in social costs between the jth regulatory alternative and the performance standard, expressed as a percentage of control costs under the latter, is given by $(1 + \alpha)(1 - S_j^*/S_0^*)$, where the ratio of social costs, S_j^*/S_0^*, can be computed using the expressions in table 7.4. Figure 7.4 plots the results as a function of m for $\alpha = 1$, where the cost coefficients are distributed independently ($V = 0$) and the sum of the variances is $\sigma^2 = 2m^2$. At $m = 1$, the two types of charges collapse to a single alternative, as do the two types of standards; the efficiency advantage of the charge over the standard is 44 percent, as it was in chapter 2. As m increases, the relative efficiency advantage of the performance charge remains constant, by design, and the efficiency of the specification charges deteriorates fairly gradually. The specification standards, however, are extremely inefficient for multistage processes; at $m = 3$, the loss in net benefits relative to the performance standard is 344 percent. At that point the optimal specification standards cause sources to spend 2.7 times more on control costs, yet leave 2.7 times more residual damage than under the optimal performance standard.

Figure 7.5 reports the results for $\alpha = 0.5$. The major qualitative differ-

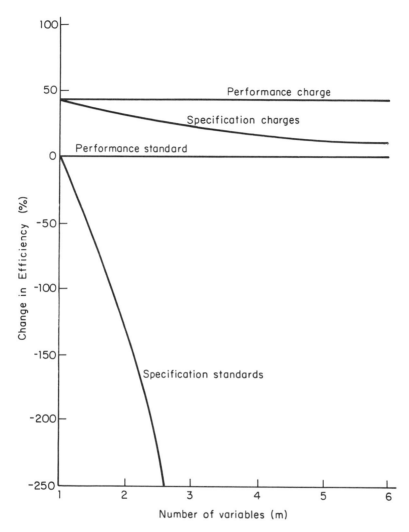

Figure 7.4   Magnitudes of efficiency changes with *m* variables ($\alpha = 1$)

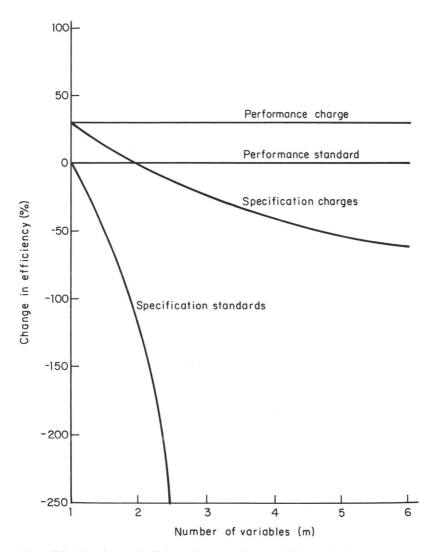

Figure 7.5   Magnitudes of efficiency changes with *m* variables ($\alpha = 0.5$)

ence from the previous figure is that the performance standard is more efficient than the specification charges for $m > 2$.

Additional Considerations

Whatever the curvature of the cost functions, the formal results suggest that a performance charge is the most efficient alternative and that specification standards are the least efficient by a large margin. As in the previous chapter, however, we need to examine some additional factors not included in the formal model to see if they affect the conclusions.

Information Requirements and Uncertainty

As with similar models in chapters 4 and 6, it is easily shown that misestimating marginal damage (λ) has the same proportional effect on the efficiencies of all of the alternatives. In particular, if λ is misestimated by a factor of k_λ, the sum of control costs and damages under each alternative rises by the factor $(k_\lambda^{\alpha\varepsilon} + \alpha k_\lambda^{-\varepsilon})\varepsilon$, as in chapters 4 and 6. Thus errors in estimating marginal damages increase the absolute efficiency differences derived previously.

Only the performance charge is insensitive to errors in estimating costs; the optimal performance charge rate depends only on marginal damage. In contrast, the performance of each of the other alternatives deteriorates if costs are misestimated. The intuition behind this result is clear for the standards-based approaches; the optimal standards depend on control costs. With the specification charges the reasoning is only slightly more complex. In the two-variable model the optimal charge on x depends on how tightly y will be controlled, and vice versa. If x is relatively expensive to control, for example, the optimum will involve relatively tight control of y and relatively little control of x. Thus the optimal charges will be higher on y than on x because, given the higher value of x, the marginal damage caused by a change in y will be greater. Perhaps somewhat counterintuitively, the lower the relative cost of controlling a variable, the higher its optimal charge. Hence errors in estimating control costs will lead to the wrong specification charges.

The specification charges, however, are considerably less sensitive than the standards to errors in estimating costs. If all of the cost coefficients are misestimated by a factor of k, it can be shown that the sum of control costs and damages with either the performance standard or the specification standard will rise by the same factor as in chapters 4 and 6, $(k^{-\alpha\varepsilon} + \alpha k^\varepsilon)\varepsilon$. Social costs under the specification charges, however, rise only by the factor $(k^{\alpha\varepsilon\eta} + \alpha k^{-\varepsilon\eta})\varepsilon$.[1] If, for example, $\alpha = 1$ (and thus $\varepsilon = 0.5$ and $\eta = 0.333$), and costs are misestimated by a factor of two, either high or low (i.e., $k = 0.5$ or $k = 2$), then social cost will rise by 6.1 percent under the standards but by less than 1 percent under the specification charges.

We see then that uncertainty strengthens the case for the performance charge and against the specification standards. Its impact on the choice between the performance standard and the specification charges is less clear-cut. Where our earlier results indicate an advantage for the specification charges, uncertainties about costs and damage both increase that advantage. Where our results under certainty indicate that the performance standard is more efficient, however, uncertainty about damages widens the gap, but uncertainty about costs narrows it.

Prices and Revenues

Of the pure "price" alternatives, only the performance charge ensures that final product prices will reflect external damages. In the two variable case the hybrid charge-standard also yields appropriate final product prices. Under the standards sources pay nothing for the residual damages they impose. Under the specification charges firms pay for the residual damages, but the amount paid by an individual source may deviate considerably from its actual damages. In fact the total charges collected will be double the residual damages (in the two variable case) because the total charges collected on each variable will be equal to total residual damages.[2] Thus, to avoid excessive effects on final prices, it might be desirable to convert the charge on *one* of the variables into a combination charge-subsidy that yielded no net revenues.

The charge-based approaches (including the hybrid charge-standard in the two variable model) offer the advantage that they generate revenues that allow reductions in other, distorting taxes. These tax-displacement benefits, as demonstrated in chapter 3, may be large.

Incentives for Rematching

We assumed in the formal model that the matches between control variables were fixed. In the case of benzene emissions from service stations, for example, this assumption would imply that each service station was matched to a particular refinery and could not switch to a different one. Rematching, however, may be both feasible and desirable.

Suppose we have two refineries, A and B, each selling to service stations that have a range of costs for controlling evaporative emissions. Reducing the benzene content of gasoline is more expensive at A than at B. We could improve efficiency by rematching service stations and refineries, shifting low-cost stations to A and high-cost stations to B. In terms of the formal model a negative correlation (ρ) between the cost coefficients reduces total costs.[3] The specification-targeted strategies provide no incentive for this rematching to occur. Under the performance standard and the performance charge, however, there is such an incentive; service stations with high control costs will be willing to pay more than other stations to purchase gasoline with a low benzene content. Thus, to the extent that this kind of rematching can go on at moderate cost, it improves the relative efficiencies of both the performance-based measures.

Summary

The qualitative considerations strengthen two basic conclusions drawn from the formal model: (1) the performance charge is substantially more efficient than any of the alternatives considered, and (2) the specification standards are a highly inefficient strategy. As in the previous chapter, we see the importance of selecting the appropriate target as well as the appropriate instrument. In contrast to the results in chapter 6, however, the choice of target is particularly critical if standards are used; the gulf in efficiencies between performance and specification-targeted strategies is large for charges, but immense for standards.

The performance charge faces two major obstacles. The first is political; Congress and regulators exhibit a strong preference for standards over charges. The second is practical; a performance target (whether the instrument is a charge or a standard) may be much more difficult and expensive to monitor than specification targets. Our findings then have two important implications: (1) if political considerations preclude the use of charges, it is worth spending a great deal extra on monitoring and other administrative costs to impose a performance standard rather than specification standards; and (2) if monitoring or other administrative problems preclude a performance target, the potential efficiency gains justify a major effort to overcome the political barriers to charges. Ideally, of course, both obstacles can be surmounted.

Chapter 8

Benzene Case Study: Overview and Damages

In June 1977 EPA listed benzene as a "hazardous air pollutant" under Section 112 of the Clean Air Act (42 *Fed. Reg.* 29332 1977). The agency's action followed the release of new epidemiological evidence linking occupational exposure to leukemia and action by OSHA to reduce the occupational exposure standard.[1] After almost three years of study EPA proposed its first benzene regulation in April 1980, an emission standard for maleic anhydride plants that use benzene as a feedstock (45 *Fed. Reg.* 26660 1980). Within nine months the agency had proposed three more regulations, one for ethylbenzene/styrene plants (45 *Fed. Reg.* 83448 1980), another for benzene storage vessels (45 *Fed. Reg.* 83952 1980), and the third covering fugitive emissions from petroleum refineries and chemical manufacturing plants (46 *Feb. Reg.* 1165 1981). As of September 1983 these proposals were still pending, and no additional benzene regulations had been proposed.

The regulation of benzene is potentially important in its own right, because benzene is a major industrial chemical, ranking among the top fifteen with a production volume of 12.7 billion pounds in 1979 (*Chemical and Engineering News* June 9, 1980, 36). Most of that output is used to produce other industrial chemicals, which in turn are used in the manufacture of a variety of products, including polyurethane foams, nylon fibers, insecticides, and reinforced plastics (Mara and Lee 1978, 21). Though not counted in production figures, roughly the same amount of benzene is contained in gasoline and other petroleum products.[2] As a result exposure is widespread, particularly in urban areas, but the potential costs of regulation also are very high.

EPA's approach to benzene is also important as a potential prototype for regulating other chemical carcinogens (leukemia is a cancer of the blood). When the agency described a "generic" policy for identifying and regulating airborne carcinogens under Section 112 (44 *Fed. Reg.* 58642 1979), it suggested benzene as an example of how that policy would be implemented (Regulatory Analysis Review Group, 1980). Benzene also has played a key role in OSHA's policy; many portions of its 1980 generic policy toward occupational carcinogens are virtually identical to its 1978 benzene standard (Nichols and Zeckhauser 1981).

Table 8.1   Estimated benzene emissions and exposure

Source category	Emissions		Exposure	
	Million pounds	Percent	Million ppb-years	Percent
Chemical manufacturing	60.0	11.0	8.5	4.7
Coke ovens	7.8	1.4	0.2	0.1
Petroleum refineries	4.1	0.8	2.5	1.4
Solvent operations	n.a.[a]	—	n.a.[a]	—
Storage and distribution of gasoline and benzene	10.2	1.9	n.a.[a]	—
Automobiles	443.6	81.5	150.0[b]	82.5
Service stations	14.6	2.7	20.6[c]	11.4
Miscellaneous	4.0	0.7	n.a.[a]	—
Total assessed	544.3	100.0	181.8	100.0

Sources: Emissions from PEDCo (1977, 1-2, 4-62). Exposure from Mara and Lee (1978, 3).
a. Not assessed.
b. Applies only to residents of SMSAs with populations greater than 500,000.
c. Composed of urban residents (19.0 million ppb-person-years) and self-service gasoline customers (1.6 million ppb-person-years).

Emissions and Exposure

After listing benzene, EPA quickly commissioned two general studies to help it set priorities for developing standards, one of emissions (PEDCo 1977) and the second of exposure (Mara and Lee 1977, 1978). Although based on very crude methods and subject to great uncertainty (and, in several cases, to demonstrable error), these estimates provide a rough idea of the relative contributions of different types of sources.[3] The results are shown in table 8.1, where exposures are summarized in part-per-billion-person-years, the product of average exposure and the number of people exposed. The estimated total exposure of 182 million ppb-person-years suggests that U.S. residents are exposed on average to ambient concentrations of less than 1 ppb. Note that automobiles account for over four-fifths of the estimated totals. Chemical manufacturing ranks second in emissions but falls to third place in estimated exposure, behind service stations. The production of benzene—which occurs primarily in petroleum refineries and, to a much smaller extent, as a by-product of coke ovens—accounts for relatively small amounts of emissions and exposure.

The levels of emissions and exposure from automobiles should fall rapidly over the next few years as the fraction of automobiles meeting existing standards increases (hydrocarbons as a class are covered). PEDCo (1977, 1–5) estimates a 75 percent reduction by 1985. The next largest source of exposure, service stations, already is subject to vapor recovery requirements (again to control hydrocarbons as a class) in most large urban areas (Mara and Lee 1978, 105). Both categories also involve large numbers of individual sources: close to 100 million automobiles and about 150,000 service stations. In contrast, the number of chemical manufacturing plants emitting benzene is relatively small; PEDCo estimated that more than half of the total was from maleic anhydride plants, of which there are less than a dozen in the whole country.[4] Thus it is not surprising that EPA placed top priority on developing an emission standard for that category. Before we can analyze the merits of specific regulatory alternatives, however, we need to examine the damages caused by exposure to benzene.

Health Effects Associated with Benzene

Benzene has long been recognized as hazardous, with reports of toxic effects dating back to the last century (U.S. EPA 1978a, 29). Until recently, however, regulatory efforts focused on preventing noncarcinogenic effects (including aplastic anemia) that are believed to occur only at concentrations far beyond those found in the ambient air. Thus, although benzene is subject to standards covering hydrocarbons as a class, EPA currently has no regulations covering it specifically. More recent evidence that benzene is a leukemogen, though also based on epidemiological studies of workers exposed to benzene at levels 1,000 times or more higher than those typically found in the ambient air, have spurred concern because many scientists believe that any exposure to a carcinogen, no matter how minute, poses a risk.

Benzene and Leukemia

Dozens of reports linking benzene and leukemia have appeared in the medical literature since 1928 (U.S. EPA 1978a, 68). Until several years ago, however, the evidence was considered inconclusive because it consisted almost entirely of clinical reports of isolated cases of leukemia in workers exposed to very high levels of benzene and, often, other potentially carcinogenic chemicals. Laboratory animal studies did not show a leukemogenic effect, and several epidemiological studies also yielded negative results. As recently as 1976 a review by the National Academy of Sciences concluded: "It is probable that all cases reported as 'leukemia associated with benzene' have resulted from exposure to rather high concentrations of benzene and other chemicals" (National Academy of Sciences 1976, ii).

EPA's and OSHA's actions in 1977 were triggered by the completion of a study by the National Institute of Occupational Safety and Health (Infante el al. 1977), which showed a much higher than expected incidence of leukemia in workers exposed to benzene while employed at two plants in the rubber industry. Within several months OSHA had issued an emergency temporary standard lowering the occupational limit tenfold, and EPA had listed benzene under Section 112.

Other epidemiological studies provide mixed support for Infante at al.'s results. Thorpe (1974) did not find a statistically higher incidence among European employees of a major oil company. Redmond et al. (1976) also found negative results among coke oven workers, even those working in areas where benzene is produced as a by-product. Ott et al. (1977) found a higher than normal incidence of leukemia among chemical workers, but the results were only of borderline statistical significance, and some of the workers (including two of the three leukemia victims) had been exposed to other suspect chemicals. Turkish shoe workers using benzene-based adhesives were found to be at greatly increased risk of leukemia and pancytopenias (Aksoy et al. 1974, 1976, 1977), as were Italian rotogravure workers using inks and solvents containing high concentrations of benzene (U.S. EPA 1978a, 74). In both of these studies cases of pancytopenia and (with a greater lag) leukemia rose after the introduction of benzene and then fell again after the benzene-based adhesives and solvents were replaced with benzene-free substitutes.

Few scientists now dispute that high-level exposures to benzene increase the risk of leukemia. The risk at the low ambient concentrations relevant to EPA's decisions, however, is hotly contested. The epidemiological data are drawn from occupational settings where exposures ranged from several to several hundred parts per million on a regular basis, with peaks exceeding 1,000 ppm. In contrast, as discussed earlier, average environmental concentrations are about one part per billion (i.e., 1,000 or more times lower). Distinguished scientists can be found on both sides of the debate, with some agreeing with EPA's position that any exposure imposes a risk, while others argue that even the current occupational limit of 10 ppm is well below the threshold needed to induce leukemia.[5]

Low-Dose Extrapolation

The problem of extrapolating from high-dose data to low-dose exposures is ubiquitous in the regulation of carcinogens. The central problem is that neither epidemiogical studies nor laboratory experiments with animals are capable of detecting low-level risks.[6] Thus, unless the chemical is a very potent carcinogen, individuals with unusually high exposures must be studied, or animals must be given doses far beyond those ever likely to be encountered by people. A variety of mathematical models has been developed to perform the necessary extrapolations. Unfortunately, current theory does not provide unambiguous support for any one of them, nor can they be selected empirically.

The "one-hit" model is the one most often used.[7] It assumes, at least metaphorically, that cancer can be induced by a single "hit" of a susceptible cell by a carcinogen.[8] Thus the lifetime risk is the probability of one or more hits. At low doses, the predicted risk is proportional to the dose; thus, for example, if the relevant dose is 1,000 times lower than that at which the risk was measured, the estimated risk is also 1,000 times lower. Because of this property it is often called the "linear" model. It is difficult to tell how much of this model's popularity is due to scientific belief in its accuracy as opposed to a value judgment that decision makers should be conservative in the face of great uncertainty; most scientists accept the linear model as providing an upper-bound estimate of the risk.

The other models commonly used are convex at low doses; as the dose is reduced, risk falls more than proportionately. Thus, when estimated from the same data, these models predict smaller low-dose risks than the linear model. The most well known of these nonlinear models is the log-probit, which was developed for the safety testing of drugs and noncarcinogenic chemicals. It assumes that individual thresholds are distributed log-normally in the population. It has two paramaters: β, which measures how rapidly the risk changes as the dose is altered, and α, which establishes the absolute level of risk.[9] The Mantel-Bryan (1961) approach, which sets β equal to unity so that only α need be estimated, is an important special case; under it each tenfold reduction in dose causes a shift of one standard deviation in the normal distribution.

Other well-known models include the logit, which yields an S-shaped curve similar to the log-probit (but with somewhat higher risks at low doses) and the multi-hit model, which is a generalization of the one-hit model.[10] At low doses, predicted risk under a "k-hit" model is proportional to the dose raised to the kth power. Thus, under a two-hit model,

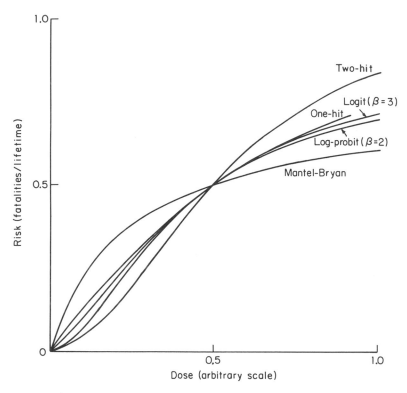

Figure 8.1   Comparison of dose-response models in the observable range

cutting the dose by a factor of 1,000 reduces the risk by a factor of $1,000^2$, or 1 million.

All of the models yield similar dose-response curves over the ranges that can be measured in laboratory and epidemiological studies. Figure 8.1 plots five such curves, where the parameters have been set to yield a risk of 0.5 at an arbitrarily scaled dose of 0.5. Note how close they are over most of the range. If we examine the models at low doses, however, the curves look very different.

Consider a hypothetical case where we have excellent data showing that continuous exposure to some chemical at 1 ppm raises the lifetime risk of cancer by 0.01. We now need to estimate the risk at 1 ppb, a thousand times lower. We fit each of the models to our data and plot the results on a logarithmic scale, with the results shown in figure 8.2. At 1 ppb the one-hit model predicts a risk of 1.0×10^{-5}. The others, which looked quite similar in figure 8.1, predict risks at least ten times lower;

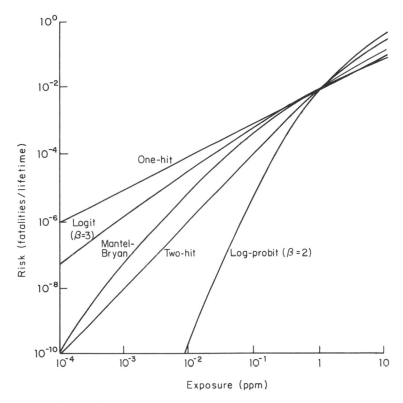

Figure 8.2 Low-dose extrapolation

predicted risk with the log-probit model ($\beta = 2$) is more than 100 *billion* times smaller.

CAG Dose-Response Function

It is impossible to derive risk estimates for benzene exposure at low levels that can be defended rigorously. EPA's (1978a, 94) review of the health effects concluded, "the data are not adequate for deriving a scientifically valid dose-response curve." EPA's Carcinogen Assessment Group, however, has estimated a dose-response function based on the linear model, three epidemiological studies, and several strong assumptions (Albert et al. 1979). That estimate, though open to many criticisms, provides a useful starting point for estimating the damage function.

In applying the linear model, the CAG made the simplifying assumption that lifetime risk is proportional to cumulative lifetime exposure; 70 years of exposure at 1 ppb poses the same risk as as 7 years at 10 ppb or 2.6 days (0.007 year) at 10 ppm (the current occupational limit). This assumption allowed the CAG to extrapolate from workers exposed to high but intermittent concentrations to the general population, exposed at much lower levels over longer periods of time. (With any of the nonlinear models this assumption clearly would be invalid, as the distribution of exposure matters, not just its mean.) The CAG also had to make some strong assumptions about the epidemiological evidence, because the exposure data were very poor and, in one case (Aksoy et al. 1974, 1976, 1977), there was considerable uncertainty about the baseline risk of leukemia in the absense of exposure to benzene.

The CAG estimated the linear coefficient separately for each of the three epidemiological studies used (Aksoy et al. 1974, Infante et al. 1977; Ott et al. 1977) and then averaged the results to get an overall estimate of 0.0241 deaths per lifetime of exposure to 1 ppm of benzene. Dividing that estimate by 1,000 (the number of ppbs per ppm) and by 70.96 (the expected years of life in the CAG life tables) yields an estimated risk of 0.339×10^{-6} deaths per ppb-person-year (Albert et al. 1979, 21). A crude estimate of the overall risk from the emissions identified by EPA may be obtained by combining this risk estimate with Mara and Lee's estimate of exposure:

$$(0.339 \times 10^{-6} \text{ deaths/ppb-yr})(181.8 \times 10^6 \text{ ppb-yr/yr})$$

$$= 61.6 \text{ deaths/yr.}$$

(8.1)

Even this estimate, which is almost certainly too high, suggests that benzene in the environment is not a major threat to public health.

The CAG has estimated the 95 percent confidence limit around its risk estimate to be 0.129×10^{-6} to 0.887×10^{-6} deaths per ppb-person-year.[11] That confidence interval, however, is based only on uncertainty about the exposure levels in the epidemiological studies; it does not include the major source of uncertainty, the method of extrapolation. Moreover several observers have questioned the CAG's implementation of the linear model, in particular its exposure estimates for all three studies, its inclusion of the deaths of workers not in the original cohort of the Infante study, its failure to exclude workers exposed to other hazardous chemicals in the Ott study, and its estimate of the baseline risk in the Aksoy study. Two EPA analysts, Luken and Miller (1979), argue that the risk estimate should be four times lower, 0.081×10^{-6} deaths per ppb-person-year. Lamm (1980), an occupational physician and consultant for the American

Petroleum Institute, argues that the CAG's estimate should be 0.0327 × 10^{-6}, ten times lower.

Valuing Risk Reduction

Benefit-cost analysis requires that the costs and benefits be expressed in the same (usually monetary) units. For airborne benzene that means that we must assign a value to "saving a life;" the nonleukemic effects of low-level exposure are almost certainly negligible and leukemia in adults is almost invariably and rapidly fatal. Over the past decade or so, a large literature, both conceptual and empirical, has grown up around the question of valuing life-saving and other health-promoting activities. A consensus appears to have emerged, at least among economists, that the value attached to reducing risk should be the standard one of willingness to pay. The principle is simple: the value of some benefit to an individual is the amount that individual would be willing to spend to secure it. A slightly different formulation asks how much an individual would have to receive in cash to be willing to forgo the benefit.[12] In its most basic form the willingness-to-pay criterion is almost tautological, but it offers some useful insights and general guidance, if not a single, unambiguous estimate.

It is important to stress that the relevant question is not how much individuals would pay to avoid certain death but rather how much to reduce the probability of premature death by some small increment. This is a question that everyone answers implicitly almost daily. Examples include the choice among automobiles with different safety records, occupational decisions, and whether one ever jaywalks to save time. At a societal level we also make these same kinds of trade-offs when we decide, for example, how much to spend on highway safety, fire protection, and environmental controls. And, much rhetoric to the contrary, it is clear that to save resources frequently we forgo opportunities to reduce risk.

Empirical Estimates
The major obstacle to implementation of the willingness-to-pay criterion is the difficulty of estimating the appropriate dollar amounts. As Schelling (1968) suggests, one approach is to ask people directly. When Acton (1973), for example, asked a small sample of individuals how much they would pay to have a mobile coronary unit that would reduce the fatality rate among heart attack victims, their answers implied a value per life

in the range $28,000 to $43,000. Few observers are comfortable with the results of such surveys, however, given the hypothetical nature of the questions and the difficulties that individuals have in dealing with issues involving small probabilities (Tversky and Kahneman 1974).

The preferred approach has been to draw inferences from actual behavior in labor markets and elsewhere, to estimate individuals' revealed preferences. Thaler and Rosen (1974) estimate that a sample of workers in high-risk occupations received $200 in extra pay for each 0.001 increase in the risk of death, implying a value per life saved of about $200,000 in 1967 dollars. Dillingham (1979) estimates a somewhat higher wage premium, about $368,000 per life for blue-collar workers in his sample. In contrast, Smith (1976) gets much higher estimates: $2.6 million for his 1967 sample and $1.5 million for his 1973 sample. Viscusi (1978) estimates similarly high premiums, in the range of $1 million to $1.5 million in 1969–70 dollars. Using data on seatbelt use, Blomquist (1977) estimates that the average driver in 1975 valued saving his life at about $257,000 (as cited in Bailey 1980, 40).

Bailey (1980) has attempted to combine these empirical estimates to derive a range of values that can be used to evaluate risk-reduction programs. His final range is $170,000 to $715,000 per life saved, with an intermediate estimate of $360,000 in 1978 dollars. These estimates are based on the studies by Thaler and Rosen, Blomquist, and Dillingham, all adjusted to achieve greater consistency in assumptions and to place them in 1978 dollars, but they do not incorporate the much higher estimates of Smith and Viscusi. Smith's 1967 figure, for example, would translate to about $5 million in 1978 dollars.

Bailey argues that the high estimates of Smith and Viscusi yield implausible predictions for individual behavior. He presents the hypothetical example of a family of four that earns $18,500 (about the median in 1978) per year and is offered the chance to reduce the annual risk of death for each of its members by 0.0005 (roughly the decline in U.S. death rates from 1970 to 1975), thus "saving" 4(0.0005) = 0.002 lives per year. Using Bailey's intermediate estimate of $360,000, the family would be willing to spend up to $720 per year, about 4 percent of its income. If it values saving a life at $5 million, it should be willing to spend $10,000 per year for this reduction, over one-half of its income, which seems highly implausible (Bailey 1980, 45–56). This example, however, does not provide a compelling argument against less extreme figures of $1 million to $2 million per life saved. Thus it is difficult to rule out values in the range from several hundred thousand to several million dollars per life saved.

Length and Quality of Life

A major difficulty with these estimates is that they fail to reflect differences in the length and quality of lives saved by different programs. Most people, for example, would spend more to save the lives of healthy thirty-year-olds than to prolong briefly the lives of seriously ill seventy-five-year-olds. To deal with this problem, Zeckhauser and Shepard (1976) argue that we should measure the benefits in terms of "quality-adjusted life-years," thus taking account of both the number of years saved and such factors as physical disability. Unfortunately, the estimates for benzene predict only the number of lives saved, and not their distribution by age. We know, however, that cancer is disproportionately a disease of the elderly. The death rate for myelogenous leukemia (the type most often associated with benzene) is 26 times higher among people 70 to 74 than among those 1 to 5 years of age (Albert et al. 1979, table 1), which suggests that the lives "saved" by reducing benzene exposure are likely to be relatively short. (Benzene has not been associated with lymphatic leukemia, the type most common in children.) Thus we might use a lower value than we would in evaluating other programs that saved a more representative cross section of the population.

Discounting also affects comparisons between programs to control carcinogens and other life-saving activities. In benefit-cost analyses, the standard procedure is to discount the streams of both costs and benefits, to reflect the opportunity cost of the funds employed and time preferences. Opinion in the economics profession is split as to whether discounting should be applied to health benefits, such as years of life saved.[13] Discounting generally makes control of carcinogens less attractive, both in absolute terms and relative to other health-related programs because of the long lag times typically associated with exposure and the onset of cancer.

Consider a simple example with two programs that both cost $100 million today. Program A saves 100 forty-year-olds from immediate death in automobile accidents, whereas program B reduces exposure to a carcinogen, thus preventing the cancer deaths of 200 forty-year-olds, twenty years from now. If we do not discount, B is clearly preferred. If we discount at a (real) rate of 5 percent, however, which most analysts would consider low, $1 received in twenty years is worth only $$1/1.05^{20} = 0.377 today, and the net present value of the benefits under B is only $(0.377)(200) = 75$ lives. Thus A should be chosen.

Critics of discounting argue that it discriminates against future citizens.[14] Consider, however, a third option, program C, which involves investing \$100 million at a 5 percent rate. In twenty years the initial investment will have grown to $(1.05^{20})(100) = \$265$ million, which may be used to implement an expanded version of A. If A can be expanded proportionally, 265 deaths from auto accidents can then be prevented. If we do not discount, C dominates both A and B. But then, why not leave the money invested for an even longer period, so that even more lives can be saved? Carried to its logical conclusion, the no-discounting position argues not for immediate expenditures on programs with delayed effects on mortality but rather for postponing programs indefinitely while invested funds grow.

The impact of discounting clearly is to reduce the benefit of controlling benzene, though the lack of data on the average lag between exposure and death from leukemia precludes an accurate quantitative estimate. A simple numerical example, however, suggests that adjusting estimates of the value of "saving a life" for lag times and for the advanced age at which cancer usually occurs may have a significant effect. Suppose we have an estimate of \$1 million, based on an immediate risk of death for a forty-year-old who otherwise could expect to live for thirty-five more years. At a 5 percent discount rate, that would imply a value of \$58,164 per year of life saved. We now wish to use that estimate to value a program that will reduce exposure to a carcinogen for that same individual, thus preventing death in fifteen years, when that person has 20 remaining years of life. If we use the implicit value per life-year, and discount at 5 percent, the value of reducing exposure is only \$366,100 per "life saved," not \$1 million.[15] Although these numbers are hypothetical, they illustrate the argument that it is appropriate to use a lower value per "live saved" in evaluating regulatory options for airborne benzene than one might use for other programs that save younger people sooner.

Summary: The Damage Function for Benzene

Massive uncertainties pervade estimates of both the risks of benzene exposure and the present value of averting a fatality due to benzene-induced leukemia. Moreover, given the nature of the uncertainties, it is difficult to develop defensible subjective probabilities to generate an expected value of the damages caused by airborne benzene. Despite these uncertainties, however, we can draw two conclusions that may provide important guidance in assessing alternative regulatory strategies.

First, the expected benefits of reducing ambient benzene concentrations probably are proportional to the reduction in total exposure. Even if one assigns only a modest probability to the linear model's being correct, the expected dose-response function will be very close to linear at the relevant doses because the other models predict near-zero risks. (If one assigns zero likelihood to the linear model, marginal damage is not constant, but then ambient benzene is not a health problem.) With constant marginal damages, as shown in chapter 4, the optimal degree of control depends critically on costs, and price-based approaches are more robust than quantity-based schemes in the face of uncertainty. Cutting exposure from 2 to 1 ppb yields the same expected reduction in damages as cutting it from 10 to 9 or from 1 to 0; there is no target level of exposure.

Second, marginal damage is almost certainly no higher than $1 per ppb-person-year and probably is much lower. Even if we combine the CAG risk estimate, which is likely to be too high, with a value per life saved as high as $3 million, the marginal benefit of reducing exposure is only $(0.339 \times 10^{-6})(\$3 \times 10^{6}) = \$1$ per ppb-person-year. More plausible estimates generate substantially smaller values. If, for example, the true risk is half that estimated by the CAG (but still double that estimated by Luken and Miller and five times higher than Lamm's estimate) and the present value of averting a future fatality due to benzene exposure is $500,000, cutting exposure by 1 ppb-person-year is worth only $0.085. With Lamm's risk estimate and Bailey's "low" value of $170,000 for each life saved, marginal benefit falls to $0.0056 per ppb-person-year, less than one cent.

Chapter 9

Benzene Case Study: Maleic Anhydride Plants

The implicit logic behind EPA's decision to make maleic anhydride plants the first category of benzene sources to be regulated was clear: it appeared that controlling a small number of plants would eliminate a relatively large quantity of emissions. Moreover within each plant virtually all of the benzene is emitted from a single point, the product-recovery absorber, for which several well-known, highly efficient control devices are available. New plants are expected to use a different feedstock, *n*-butane, regardless of EPA's actions, because higher oil prices have driven up the cost of benzene. As a result the portion of the proposed standard that forbids any benzene emissions from new plants has no apparent costs or benefits. The analysis here is restricted to existing plants.

The proposal for existing plants is a performance standard: emissions from the product recovery absorber are not to exceed 0.3 kg benzene for each 100 kg benzene input. No particular control method is required, but firms would have to monitor emissions continuously to show compliance (45 *Fed. Reg.* 26660 1980). The proposal is commendable for its emphasis on performance and for the extensive analysis that preceded it. As noted in chapter 1, however, it suffers from the inevitable faults of any uniform standard; it ignores variations in both the marginal costs and the marginal benefits of controlling emissions. Moreover the level of control chosen is difficult to justify under benefit-cost criteria. Thus it is worthwhile to consider several alternative strategies, in particular those analyzed on a theoretical level in chapter 6: a uniform standard, a standard conditional on exposure levels, a uniform emission charge, and a uniform exposure charge. (The data are insufficient to consider the fifth option from chapter 6, a uniform exposure standard.)

Most of the analysis in this chapter is based on information available to EPA before it proposed the standard in April 1980. The intent is to illustrate what the agency can do using data it already collects. At the end of the chapter, new information that became available after the proposal was made is introduced and used to test how strategies based on the earlier, incomplete information would perform.

Overview of the Maleic Anhydride Industry

EPA estimated that maleic anhydride was produced in ten plants in the United States, with a total annual capacity of 229 million kg (U.S. EPA 1980, 1-3). Three production methods are used: oxidation of benzene, oxidation of n-butane, and recovery of maleic anhydride as a by-product of the manufacture of phthalic anhydride. These processes account, respectively, for 82, 16, and 2 percent of total capacity (U.S. EPA 1980, 1-3). Slightly over one-half of production is employed in the manufacture of unsaturated polyester resins, which in turn are used in reinforced plastics. Other major uses of maleic anhydride include agricultural chemicals, lubricating additives, and fumaric acid (U.S. EPA 1980, 5-8).

Eight of the ten plants use the benzene oxidation process and thus would be subject to the regulation.[1] In those plants the gaseous product of the oxidation process is routed through a product-recovery absorber, where about 99 percent of the plant's total benzene emissions are released through a vent. In the absence of control devices the emission rate depends primarily on the conversion rate of benzene in the production process. Estimated plant-specific conversion rates vary from 90 to 97 percent, but for analytic purposes EPA has assumed a uniform conversion rate of 94.5 percent, which results in an uncontrolled emission rate of 0.067 kg benzene per kilogram maleic anhydride. Benzene accounts for almost four-fifths of the total emissions of volatile organic chemicals from benzene-fed maleic anhydride plants, with formaldehyde, maleic acid, and formic acid comprising the remainder (U.S. EPA 1980, 1-6 to 1-12).

Table 9.1 presents some basic information on the eight benzene-fed plants. Five of them already had control devices on their absorber vents as of 1980, primarily in response to state hydrocarbon regulations. The two Reichold plants use carbon adsorbers, where the exhaust from the vent is passed through beds of carbon for adsorption and recovery. The Denka and Koppers plants use thermal incineration, combusting the exhaust at high temperatures. One plant, U.S. Steel, uses catalytic incineration, which employs a catalyst to permit destruction of the hydrocarbons at relatively low temperatures. Table 9.1 shows estimated emission levels at full capacity operation, both uncontrolled and with current devices in operation. In addition to the three types of control in current use, plants can reduce benzene emissions by altering the production process, or eliminate them altogether by switching to an alternative feedstock.

Table 9.1   Basic data on benzene-fed maleic anhydride plants

Plant	Location	Capacity (1 million kg/yr)	Current Control Level (%)	Benzene emissions (1,000 kg/yr)[a] Uncontrolled	Current
Ashland	Neal, W.V.	27.2	0	1,821	1,821
Denka	Houston, Tex.	22.7	97	1,520	46
Koppers	Bridgeville, Pa.	15.4	99	1,031	10
Monsanto	St. Louis, Mo.	38.1[b]	0	2,551	2,551
Reichold	Morris, Ill.	20.0	90	1,339	134
Reichold	Elizabeth, N.J.	13.6	97	911	27
Tenneco	Fords, N.J.	11.8	0	790	790
U.S. Steel	Neville Island, Pa.	38.5	90	2,578	258
Total		187.3		12,541	5,637

Source: U.S. EPA (1980, tables 1-2, 1-5) and author's calculations.
a. All estimates assume full-capacity operation and uncontrolled emission rate of 0.067 kg benzene for each kilogram of maleic anhydride.
b. Capacity shown for Monsanto reflects only portion that uses benzene feedstock. Total capacity is 48 million kilograms.

Plant-Specific Costs and Benefits of Control

Before issuing its proposed standard, EPA seriously considered only two alternative levels: 97 and 99 percent control.[2] These high levels of control were dictated, apparently, by the agency's interpretation of Section 112 as requiring the "best available technology." Moreover at each control level EPA analyzed only two control techniques: carbon adsorption and thermal incineration. Process modifications and catalytic incineration were eliminated as unable to meet the desired control levels. EPA felt itself unable to analyze feedstock substitution, as much of the data on the *n*-butane process were proprietary and there was great uncertainty about the costs of converting existing plants (U.S. EPA 1980, 1-11). Because of these limitations, our analysis of alternative strategies must be crude.

Control Costs
Ideally, we would like detailed plant-specific cost estimates for each alternative control level. Unfortunately, even when the number of plants is small, such analyses cannot be performed because they would be too costly and would require access to proprietary information. EPA's method with maleic anhydride plants was typical of its approach to cost estimation generally: a contractor developed a hypothetical model plant and per-

Table 9.2   Control costs

Plant	Control costs ($1,000/yr)[a]	
	97 percent standard	99 percent standard
Ashland	410	425
Denka	0[b]	369
Koppers	0[b]	0[c]
Monsanto	542	560
Reichold (Ill.)	320	333
Reichold (N.J.)	0[b]	245
Tenneco	213	224
U.S. Steel	547	565
Total	2,032	2,721

Source: U.S. EPA (1980, tables 5-5a to
5-6a).
a. All cost estimates assume full-capacity
operation.
b. Plant already meets 97 percent standard.
c. Plant already meets 99 percent standard.

formed engineering analyses to determine what equipment would be needed to meet the control levels. Capital costs were estimated, amortized (at 10 percent over ten years), and combined with estimated operating costs to predict total annualized costs for the model plant. The contractor then estimated the costs for each existing plant by adjusting the model-plant figures for capacity. Table 9.2 reports the results for carbon adsorption, which had slightly lower estimated costs than incineration.[3]

The cost estimates, which total $2.0 million annually for 97 percent control and $2.7 million for 99 percent control, assume full-capacity operation. Net annualized costs are slightly higher at lower levels of operation; the small savings in operating costs are more than offset by the reduction in the credit for the benzene recovered.[4] Note that as three of the plants already meet or exceed 97 percent control, they would not incur any costs beyond monitoring under the proposed standard. One plant, Koppers, would not incur any additional costs even at the 99 percent control level. EPA has assumed, however, that if a plant's existing controls do not meet a standard, its costs would be the same as for a plant currently without any emission controls. That seems unduly pessimistic; some of the equipment now in use might be adapted to tighter control levels, and at the very least operating costs would be reduced if existing devices were shut down.

Table 9.3   Reductions in emissions and exposure

Plant	Emissions reduction (1,000 kg/yr)		Exposure factor (ppb-yr/kg)	Exposure reduction (1,000 ppb-yr)	
	97 percent standard	99 percent standard		97 percent standard	99 percent standard
Ashland	1,722	1,788	0.017	29	30
Denka	0	18	0.248	0	4.5
Koppers	0	0	0.162	0	0
Monsanto	2,412	2,505	0.391	943	979
Reichold (Ill.)	61	110	0.008	0.5	0.9
Reichold (N.J.)	0	11	0.384	0	4.1
Tenneco	747	776	0.195	146	151
U.S. Steel	117	211	0.179	21	38
Total	5,059	5,418		1,139	1,208

Sources: U.S. EPA (1980) and author's calculations.

Benefits of Control

Table 9.3 reports estimated plant-specific reductions in emissions and exposures at the two control levels. The emissions reductions are measured relative to "current" levels from the previous table; thus none is shown for Denka, Koppers, or Reichold (N.J.) at 97 percent, or for Koppers at 99 percent. The center column displays estimated exposure factors, which are based on plant-specific population data, but the general dispersion modeling was not adjusted for differences in meteorological conditions or for other plant-specific parameters, such as stack height and exit velocity.[5] The figures in the two final columns, reductions in exposures, are simply the products of the exposure factors and the relevant reductions in emissions.

Two interesting facts emerge from table 9.3. First, marginal damages vary widely across plants, with the exposure factors ranging from 0.008 ppb-person-years for each kilogram emitted from the Reichold plant in rural Illinois to a high of almost 0.4 for the Monsanto plant in urban St. Louis. Second, a single plant, Monsanto, accounts for over 80 percent of the estimated benefits at either level of control.

Cost Effectiveness of Regulatory Alternatives

Table 9.4 summarizes the effects of the two standards considered by EPA. The entries under "annual costs and benefits" are simply the totals from tables 9.2 and 9.3. The cost-effectiveness ratios are the costs divided by the relevant benefits. EPA's proposed standard, 97 percent, has a ratio of

Table 9.4 Cost effectiveness of uniform emission standards

	Change in control level		
	Current to 97 percent	Current to 99 percent	97 to 99 percent
Annual costs and benefits			
Control cost ($1,000)	2,032	2,721	689
Reduced emissions (1,000 kg)	5,059	5,418	359
Reduced exposure (1,000 ppb-yr)	1,139	1,208	69
Cost effectiveness			
Emissions ($/kg)	0.40	0.50	1.92
Exposure ($/ppb-yr)	1.78	2.25	10.0

$1.78 per ppb-person-year of exposure reduction; that is, for the proposed standard to yield positive net benefits, the benefit per unit of exposure reduction, V, would have to exceed $1.78, which is well above the plausible range derived in the previous chapter. (Even with the CAG risk estimate one would have to value "saving a life" at $5.3 million to justify the standard.) The *average* ratio for the 99 percent standard is only slightly higher, $2.25 per ppb-person-year, but the marginal ratio, which is the relevant one for decision-making purposes, is much higher: dividing the incremental costs of the tighter standard by the incremental reduction in exposure yields a cost of $10 per ppb-person-year, ten times higher than the upper-bound estimate of the marginal benefit in chapter 8.

These ratios assume full-capacity operation. Adjusting the estimates for lower levels of utilization increases the costs slightly, as noted earlier, and reduces the benefits. At 56 percent utilization (the average in 1977), for example, costs for the 97 percent option rise to $2.1 million and the reduction in emissions falls to 641,000 ppb-person-years, for a cost-effectiveness ratio of $3.32.

Improving the Uniform Standard
The high cost per unit of benefit under the proposed standard is due in part to the fact that of the five plants that would need new control equipment to meet it, two already achieve 90 percent control. Relaxing the standard to that level would allow those plants to use their existing controls and would save a considerable amount of money with little change in benefits. Unfortunately, EPA has not developed cost estimates for 90 percent controls. We can make a conservative estimate of the net benefits of relaxing the standard by assuming that 90 percent controls would cost just as much as those achieving 97 percent for the three plants that cur-

Table 9.5   Cost effectiveness of less stringent standard

	Change in control level	
	Current to 90 percent	90 to 97 percent
Annual costs and benefits		
Control cost ($1,000)	1,165	867
Reduced emissions (1,000 kg)	4,646	413
Reduced exposure (1,000 ppb-yr)	1,064	75
Cost effectiveness		
Emissions ($/kg)	0.25	2.10
Exposure ($/ppb-yr)	1.10	11.5

rently are uncontrolled. That is, we assume that with a less stringent standard, Ashland, Monsanto, and Tenneco would cut emissions by only 90 percent, but their costs would be the same as at 97 percent. Table 9.5 reports the results (assuming full-capacity operation). The estimated exposure reduction is only 6.4 percent lower than at 97 percent, but costs fall 42.7 percent; the cost-effectiveness ratio drops to $1.10 per ppb-person-year of exposure reduction. More dramatically, to justify 97 percent, V must be greater than $11.[6]

Conditional Standard
The conditional standard, as explained in chapter 6, improves cost effectiveness by setting differential control levels based on differences in marginal damages. In extreme form it leads to plant-specific standards, as each plant has a different exposure factor. A more practical approach is to establish a limited number of categories; in this case, with only eight plants, it is difficult to justify more than two classes. Let us split them evenly, based on the exposure factors in table 9.3: four high-exposure plants [Monsanto, Reichold (N.J.), Denka, and Tenneco] and four low-exposure plants [Koppers, U.S. Steel, Ashland, and Reichold (Ill.)]. Table 9.6 presents the results for each group. In the high-exposure group 97 percent is justified if V exceeds $0.69 (90 percent has a higher cost-effectiveness ratio), and tightening the standard to 99 percent yields positive net benefits only if the marginal benefit of controlling exposure exceeds $12.70. In the low-exposure group even 90 percent control requires a V of almost $15. Imposing 97 percent control only on the high-exposure plants, with no new controls on the other plants, yields 96 percent of the benefits of the proposed uniform standard at 37 percent of its cost. That conditional standard dominates the uniform 90 percent

Table 9.6 Cost effectiveness of conditional standard

	High exposure		Low exposure	
	97 percent	99 percent	90 percent	99 percent
Annual costs and benefits				
Control cost ($1,000)	755	1,398	410	1,323
Reduced emissions (1,000 kg)	3,007	3,159	1,639	2,109
Reduced exposure (1,000 ppb-yr)	1,089	1,139	28	69
Marginal Cost Effectiveness				
Emissions ($/kg)	0.24	4.26	0.25	1.94
Exposure ($/ppb-yr)	0.69	12.70	14.89	22.18

alternative, achieving slightly greater benefits at 71 percent of its cost. Thus even a crude, two-level approach significantly improves the cost effectiveness of standards here.

Uniform Emission Charge

Whereas the conditional standard deals with variability in marginal damages, a uniform emission charge copes with differences in the marginal costs of controlling emissions. As shown in chapter 6, such a charge will not be fully efficient when plant-specific marginal damages vary, as they do here, but it is likely to be an improvement over a uniform standard. Under a uniform emission charge, as the charge is raised, firms will adopt controls in order of their cost per unit of emissions reduction. Table 9.7 ranks plant-control combinations in that order and also shows the ranking by marginal cost per unit of exposure reduction.[7] The two rankings are similar but not identical. Because it is the second-largest plant without controls, Ashland ranks second in cost per unit of emissions controlled, just behind Monsanto. Ashland, however, has a low exposure factor, so it ranks lower in exposure cost effectiveness; its marginal cost per unit of exposure reduction is almost 25 times higher than Monsanto's.

Computing the appropriate uniform emission charge is not straightforward. If the value per unit of exposure reduction is V and the exposure factor is E, then the exposure charge should be VE. The difficulty here is that E varies across plants. Which value should we use? One solution is to use the average value for the plants covered by the regulation. The simple average of the factors in table 9.3 is 0.198 ppb-person-years for each kilogram emitted. If we weight the factors by plant capacities, the average is 0.203. For simplicity, let us use a value of 0.2, so that the tax per kilogram emitted should be $0.2V$. If we use our high estimate of the marginal value of exposure reduction, $V = \$1$, for example, then the uniform emission

Table 9.7   Marginal cost-effectiveness rankings

Plant	Control (%)	Emissions Rank	Emissions ($/kg)	Exposure Rank	Exposure ($/ppb-yr)
Monsanto	99	1	0.224	1	0.57
Ashland	99	2	0.238	4	14.1
Tenneco	97	3	0.285	2	1.46
Tenneco	99	4	0.379	3	1.95
U.S. Steel	99	5	2.68	5	15.0
Reichold (Ill.)	99	6	3.04	8	374.
Denka	99	7	20.5	7	82.7
Reichold (N.J.)	99	8	22.7	6	59.1

charge would be $0.20. According to table 9.7, however, the lowest cost-effectiveness ratio is $0.224 per kilogram so this approach would not lead to any control unless V exceeded $0.224/0.2 = 1.12 per ppb-person-year, at which point Monsanto would control at 99 percent. For $V > 0.238/0.2 = 1.19, Ashland would also control at 99 percent. Tenneco would add controls at the 97 percent level for $V > 0.285/0.2 = 1.43 and at 99 percent for $V > 0.379/0.2 = 1.90. With all three plants controlled at 99 percent, costs would be 41 percent lower than with EPA's proposed uniform standard and the reduction in exposure would be slightly higher.

These results, though a substantial improvement over the uniform 97 percent standard, clearly are suboptimal. Under this averaging approach the emission charge is not high enough to induce Monsanto to control unless $V > 1.12, although controls at that plant yield positive net benefits if $V > 0.57. Conversely, the average emission charge leads Ashland to control if $V > 1.19, when in fact such control is not justified unless the marginal benefit of reducing exposure exceeds $14. These anomalies suggest a somewhat more sophisticated approach. Table 9.7 indicates that, for $V < 0.57, the emission charge should be low enough that no plant adds new controls. For larger values of V, the charge should induce Monsanto to control, but not Ashland. If only Monsanto controlled (at 99 percent), 86 percent of the benefits of a 97 percent standard would be realized at only 28 percent of the cost. The difficulty is that the emission cost-effectiveness ratios for the two plants are so close: $0.224 per kilogram for Monsanto and $0.238 per kilogram for Ashland, a difference of about 6 percent. Thus it would be almost impossible to set an emission charge that was precisely right.

For higher values of V, it is never cost beneficial to get Ashland to control without also inducing Tenneco to do so. Adding Ashland and Tenneco at 99 percent reduces exposure by an additional 181,000 ppb-person-years at an incremental cost of $649,000, for a cost-effectiveness ratio of $3.60. Thus, only if $V > \$3.60$, would it be desirable to set the emission charge high enough to induce any plant other than Monsanto to control.

Exposure Charge
A uniform exposure charge, as shown in chapter 6, takes account of variations in both marginal costs and marginal damages, thus combining the virtues of an emission charge and a conditional standard. Although exposure levels cannot be measured directly, they could be estimated for maleic anhydride plants by using measured emissions and plant-specific exposure factors, such as those in table 9.3. The optimal exposure charge is equal to the marginal damage per unit of exposure, V. The exposure cost-effectiveness ranking in table 9.7 gives the predicted order in which plants would add controls as the exposure charge were raised. These ratios suggest that, if $V < \$0.57$, no plants would add 97 or 99 percent controls. For $\$0.57 < V < \$1.46$, only Monsanto would control, yielding 86 percent of the estimated benefits of EPA's proposal at less than 28 percent of the cost. Raising the charge to $1.46 would, according to the estimates in table 9.7, induce Tenneco to add 97 percent controls. For $V > \$1.96$, Tenneco is predicted to install 99 percent controls. With both Monsanto and Tenneco at 99 percent control, over 99 percent of the benefits of the uniform 97 percent standard would be achieved at less than 39 percent of the cost. To induce plants other than Monsanto and Tenneco to control, the exposure charge would have to be in excess of $14 per ppb-person-year.

Net Benefits
Our analysis suggests that all of the alternatives examined would be more cost-effective than EPA's proposal. For any particular estimate of the unit benefit of controlling exposure (V), the estimated net benefit of an alternative is simply its reduction in exposure times V, minus control costs. With the exception of EPA's proposal, the controls at each plant, and hence the costs and exposure reductions, also depend on V. Figure 9.1 presents the predictions of which plants would add controls under each alternative strategy for values of V up to $5 per ppb-person-year. The first column of table 9.8 shows the estimated net benefits under EPA's proposed 97 percent standard for selected values of V in that same range. The other columns show the *change* in net benefits under each of the alternatives. Over the whole range all outperform the EPA proposal by at least $491,000 per year. The exposure charge's advantage never falls below

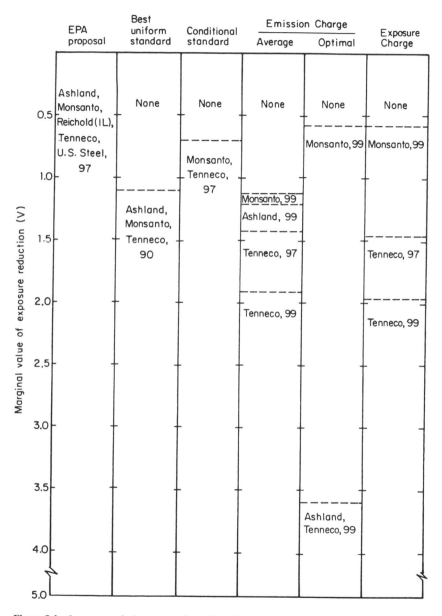

Figure 9.1   Incremental plant-control combinations under alternative strategies

Table 9.8   Net savings under alternative strategies (in $1,000/yr)

V ($/ppb-yr)	Net benefits of EPA standard	Net gains under alternatives				
		Optimal uniform standard	Condi- tional standard	Average emission charge	Optimal emission charge	Exposure charge
0	−2,032	2,032	2,032	2,032	2,032	2,032
0.75	−1,178	1,178	1,240	1,178	1,353	1,353
1.00	−893	893	1,227	893	1,312	1,312
1.50	−323	754	1,201	855	1,232	1,235
2.00	+246	717	1,177	867	1,153	1,232
3.00	+1,385	642	1,126	888	993	1,223
4.00	+2,524	567	1,076	910	834	1,215
5.00	+3,664	491	1,025	931	673	1,206

$1.2 million annually. The optimal emission charge does almost as well, but the reader should remember that its performance is extremely sensitive to the accuracy of the cost estimates, as the difference between the costs per unit of emissions controlled is very small for Monsanto and Ashland.

These results suggest that EPA's proposed standard would yield negative net benefits for any plausible estimates of the marginal benefit of reducing exposure to benzene. Indeed, although the alternative strategies do better than EPA's, none is likely to yield net gains over the status quo unless the value of reducing exposure is near the high end of the plausible range.

Robustness under Uncertainty

Our conclusions must be extremely tentative given major uncertainties in both the cost and benefit estimates. These uncertainties suggest the need to analyze how the various strategies would be affected by errors in the estimates, to perform the type of analysis conducted at a theoretical level in chapter 4. As all of the strategies are sensitive to errors in estimating damages, the focus will be on uncertainty about the costs.

The standard-based strategies are most sensitive to the accuracy of the cost estimates. The efficiency of the uniform standard depends on the accuracy of the estimate of total costs; the conditional standard also requires cost estimates for each exposure subgroup.

Contrary perhaps to theoretical expectations, the performance of the best uniform emission charge also is dependent on the accuracy of the cost estimates. Recall that the results in table 9.8 assume that it is possible to establish an emission charge that will induce Monsanto, but not Ashland, to control. If the estimates are accurate, a charge of $0.23 per kilogram would do so. Suppose, however, that control costs were 10 percent lower than estimated by EPA; a charge of $0.23 would then induce Ashland as well as Monsanto to control, although its cost per unit of exposure reduction would still be in excess of $12. Conversely, suppose costs were 10 percent higher than estimated; neither plant would control in response to the emission charge, although control at Monsanto might still yield positive net benefits. As actual costs are likely to vary from the EPA estimates by at least 10 percent, the performance of the "optimal" emission charge is shaky indeed.

In contrast, the optimal exposure charge is independent of the costs. Changes in costs may lead to other than expected levels of control, but those outcomes will not be inefficient. Suppose an exposure charge of $1 per ppb-person-year were imposed. With the EPA estimates we predict that only Monsanto will control, yielding net benefits of ($1 per ppb-person-year) (979,000 ppb-person-years) − $560,000 = $419,000. Now suppose that actual costs were half the EPA estimates, so that the charge of $1 per ppb-person-year would also lead Tenneco to control at 99 percent. Control efforts would be greater than predicted, but given the reduction in control costs that would be optimal, net benefits would rise to ($1) (979,000 + 151,000) − (0.5) ($560,000 + $224,000) = $738,000 per year. With expected benefits proportional to exposure, the efficient level of the exposure charge does not change even if costs are radically different than predicted.

Impact of New Information
New data that became available to EPA after the standard was proposed allow us to perform a useful thought experiment: suppose EPA had imposed one of the regulatory alternatives in April 1980 based on the information it had then. How well would that alternative have performed in light of the information that was presented to EPA by the Chemical Manufacturers' Association in August of that year?[8]

The most startling news offered by the CMA was that two of the three uncontrolled plants, Monsanto and Tenneco, would install 97 percent controls (thermal incinerators) whether or not EPA promulgated the standard, because of formal rules or informal pressures from state regulators. The effect of these new controls is to eliminate virtually all of the benefits of the proposed standard and much of its cost. The CMA also announced, however, that EPA had overlooked a maleic anhydride plant

Table 9.9   Cost effectiveness of proposed standard, original and revised data

	Original data	Revised data
Annual costs and benefits		
Control costs ($1,000)	2,032	1,476
Reduced emissions (1,000 kg)	5,059	2,578
Reduced exposure (1,000 ppb-yr)	1,139	62
Cost effectiveness		
Emissions ($/kg)	0.40	0.57
Exposure ($/ppb-yr)	1.78	23.96

—a small, uncontrolled facility operated by Pfizer in Terre Haute, Indiana.[9]

The net effect of these changes is to worsen substantially the cost effectiveness of the EPA proposal. Monsanto and Tenneco were the only uncontrolled plants located in relatively densely populated areas, and the Pfizer plant has an estimated exposure factor of only 0.016 ppb-person-years per kilogram, lower than that of any plant other than Reichold (Ill.). The estimated cost for each unit of exposure reduction at that plant is almost $18.[10]

Table 9.9 compares the original estimates of the costs and benefits of the proposed 97 percent standard (from table 9.4) with the revised estimates, after Monsanto and Tenneco have been eliminated and Pfizer added. Only four plants would be affected by the standard. Three have very low exposure factors (and one of them already achieves 90 percent control). The fourth, U.S. Steel, has a moderate exposure factor, but it currently controls 90 percent of its emissions, so control expenditures buy little net reduction in damages. As a result total estimated costs of the standard are reduced by 27 percent, but exposure-reduction benefits virtually disappear; the cost-effectiveness ratio increases by more than a factor of 13, from under $2 to almost $24 per ppb-person-year. Note that a more conventional analysis that focused on emissions reduction as the summary measure of benefits would grossly understate the true significance of this new information because the cost per kilogram controlled rises only 42.5 percent.

The CMA also disputed EPA's cost estimates. It asked each of the firms involved to estimate the capital cost of controls. (In apparent contradiction to EPA's relative cost estimates, all of the firms indicated that they would use thermal incineration rather than carbon adsorption.) CMA then annualized those figures using EPA's assumptions and added EPA's

Table 9.10   Comparison of EPA and CMA estimates of annualized costs

Plant	EPA estimate ($1,000)	CMA estimate ($1,000)	Percent increase
Ashland	410	1,940	373
Pfizer	199	590	196
Reichold (Ill.)	320	880	175
U.S. Steel	547	2,070	278
Total	1,476	5,480	271

Sources: EPA estimates from U.S. EPA (1980). Pfizer interpolated by author. CMA estimates from CMA (1980).

own estimates of operating costs to obtain total annualized operating costs (CMA 1980). Table 9.10 shows the results. For the four plants that would be affected by the standard, the CMA estimated total is $5.5 million, nearly four times higher than the figure in table 9.9. If we substitute the CMA cost estimate, the cost-effectiveness ratio rises to $88 per ppb-person-year.

Even with the original EPA estimates, the proposed standard was difficult to justify on benefit-cost grounds. With the new information, even if one discounts entirely the CMA cost estimates, the task seems impossible. Moreover additional controls cannot be justified at any individual plant. Using the EPA cost estimates, Ashland is the most cost-effective candidate, but its cost per ppb-person-year is over $14. With the CAG estimates Pfizer is the best plant to control, but its ratio is $53 per ppb-person-year. Thus it appears that, for any plausible value of V, the optimal regulatory strategy can do no better than the status quo.

Impact on Alternative Strategies

Let us now consider the impact of the new information on the alternative strategies. Suppose, for example, that EPA had chosen its strategy using a value of $2 per ppb-person-year. (The analysis in chapter 8 suggests that value is much too high, but it suits the illustration because it yields positive net benefits for the EPA proposal using the original estimates.) As shown earlier in figure 9.1, the best uniform emissions standard would have been 90 percent, yielding net benefits of $963,000, a gain of $717,000 over the EPA proposal (table 9.8). With the new information, however, the 90 percent standard would affect only Ashland and Pfizer. Exposure would be reduced by only 40,100 ppb-person-years, at a cost of $609,000 using the EPA estimates. Net benefits, again using $V = \$2$, would be

($2) (40,100) − $609,000 = −528,800; incomplete information would have cost $528,800, relative to what could have been achieved with knowledge about the controls at Monsanto and Tenneco and about the existence of Pfizer. If the CMA cost estimates are correct, the net loss would be much higher, over $2.4 million annually.

Suppose instead that EPA had used a conditional standard. As shown in figure 9.1, the optimal two-part standard would have been 97 percent for high-exposure plants and none for the others. With the new information, however, such a standard has no effect, as the four plants with less than 97 percent control are all in the low-exposure category. Thus the incorrect information would not have led EPA to set the wrong conditional standard in this case.

Now consider the three charge schemes. Under the average emission charge, $V = \$2$ translates to a charge of $0.40 per kilogram. The original prediction was that Monsanto, Tenneco, and Ashland would control at 99 percent; now, Ashland and Pfizer control. Exposure is reduced by 41,100 ppb-person-years at a cost of $624,000 according to the EPA figures; net benefits are ($2) (41,100) − $624,000 = −$541,800. Perhaps counterintuitively, the average exposure charge does better if the CMA cost estimates are correct; in that case none of the plants finds control attractive, and maximum possible net benefits ($0) are reaped.

With the original data the optimal emission charge for $V = \$2$ was about $0.23 per kilogram, with the expectation that only Monsanto would control. With the new information, however, such a charge is not predicted to lead to any new controls, regardless of whether the EPA or CMA cost estimates are correct. The exposure charge also leads to the optimal outcome of no new controls; a charge of $2 per ppb-person-year is far below the cost-effectiveness ratio for any of the remaining plants.

The robustness of the exposure charge should not come as any surprise, of course. In this example the optimal emission charge and the conditional standard also respond well to the new information. Unlike the exposure charge, however, those strategies will not always do so well in the face of uncertainty about costs and existing controls. Suppose that Pfizer were located in a high-exposure area but already had achieved 90 percent control; the conditional standard would have forced it to install 97 percent controls, with negative net benefits. Conversely, suppose that Pfizer were located where it is, with a low-exposure factor, but that it had slightly lower costs than the EPA estimates; the optimal emission charge would have led it to control, though in fact such control would be suboptimal given its low marginal damages. Hypotheticals such as these suggest that the apparent robustness of the conditional standard and the optimal emission charge in this instance is due more to luck than to inherent virtue.

Postscript

As of September 1983 the final standard for maleic anhydride plants had not been promulgated, although EPA officials expected a decision soon on whether to promulgate or withdraw the proposal. After the public hearing in August 1980, several changes occurred that made the proposed standard even less cost effective and further demonstrated the need for strategies that are flexible in response to shifting conditions.[12] Two more plants—Tenneco and Reichold (Ill.)—closed. Two other plants—Ashland and Denka—converted to n-butane. Monsanto installed controls and began to convert all of its capacity to n-butane. EPA also reduced the proposed control level to 90 percent, so that the standard, if ever promulgated, would apply to a single, small plant (Pfizer) located in a lightly populated area. Using EPA's exposure modeling and the CAG risk estimate, a fatality would be prevented about once every 300 years.

Chapter 10

The Role of Targets and Instruments in Reform

If reform of environmental regulation is to achieve its full potential, it must alter not only the instruments employed but also the targets to which they are applied. The importance of making regulation sensitive to cost differences has long been recognized and now is beginning to be reflected in practice. The need to be sensitive to variations in benefits has received little attention but is also a critical ingredient in making regulation more efficient. Students of environmental regulation and regulators themselves need to shift their conceptual foci from reducing emissions to reducing damages. Once stated, the point that emissions matter only insofar as they cause damages seems obvious, even trivial. But it is rarely recognized, much less reflected in the design of regulations. This shift of focus is not a mere theoretical nicety—it implies the need for major changes in regulatory strategies, whether they employ the incentives favored by most economists or the standards favored by most legislators and regulators.

The Case for Exposure Charges

The case study of maleic anhydride plants and the theoretical analyses of earlier chapters both argue that efficiency can best be achieved by employing economic incentives and aiming them at a target as closely related to damages as possible. For most health-threatening environmental hazards it appears that exposure offers an excellent target. The choice of instrument can be narrowed to charges and marketable permits (preferably allocated by auction rather than free of cost); subsidies distort the prices of final goods and require that taxes be raised to finance them. If the expected marginal damage caused by exposure varies relatively little with ambient concentrations, as appears to be the case with benzene and other airborne carcinogens, charges are preferable to permits because of their ability to adjust control levels automatically in response to changes in control costs.

The relative importance of taking account of variations in marginal damages is growing. During the 1970s most expenditures for control of air pollution from stationary sources were made by existing sources that were not subject to uniform national emission standards but rather to state implementation plans (SIPs) designed to achieve ambient standards for specified pollutants. Many states adopted very crude strategies that in essence required uniform percentage reductions in emissions across industries. As a result the marginal costs of controlling emissions varied widely, but the variation in marginal damages was limited by the fact that each SIP applies only to a limited geographic area. Across states, control efforts varied very roughly in relationship to marginal damages because the states needing the largest proportional reductions typically were those with dense urban areas.

New sources, and existing sources of hazardous air pollutants such as benzene, however, are subject to uniform national regulations. These regulations are written separately for specific categories of facilities (in this case, maleic anhydride plants), so the variability in marginal control costs, though still present, is limited. The variation in marginal damages, however, can be enormous because of the wide geographic distribution of sources subject to each regulation. The 50-fold difference in exposure factors found with maleic anhydride plants is by no means unique. Estimated exposure factors for service stations and automobiles vary by factors of 150 or more between large and small urban areas (Nichols 1981, ch. 11). Exposure factors for coke ovens, another category of sources being considered for regulation under Section 112 of the Clean Air Act, also vary more than 100-fold (Haigh 1982a). Switching from uniform standards to uniform incentives targeted on emissions would miss a large fraction of the potential efficiency gains in such cases.

Compared to a uniform emission standard, an exposure charge

1. allows sources to employ the least-cost mix of actions, including site selection, to reduce damages;

2. allocates control efforts efficiently across sources, taking into account variations in both marginal damages and marginal costs;

3. causes residual damages to be reflected in final product prices and provides appropriate entry and exit signals for firms;

4. generates revenues for the government that can be used to cut tax rates, thus reducing deadweight losses;

5. is robust in the face of uncertainty about control costs, and its performance is no more sensitive than that of the standard to uncertainty about damages;

6. responds efficiently to changes in costs over time and provides a continuing incentive for firms to develop innovative techniques for reducing damages;

7. promotes equity by charging firms for the external damages they impose and placing the same implicit value on protecting individuals regardless of where they live.

Emission standards that vary with population densities, uniform emission charges, and marketable permits denominated in exposure units offer some of these advantages, but none provides all of them.

Regulators are likely to raise several important questions about the practicality of exposure charges: (1) How should charges be set in the face of major uncertainties about the risks posed by pollutants and disagreement about the dollar value that should be placed on reducing risk? (2) Would exposure charges require considerably more information than now collected for standards, thus prolonging the regulatory process and increasing its expense? (3) Could compliance be monitored and enforced at reasonable cost? (4) What should be done with the revenues raised, which may be substantial? (5) What assurance could be provided that environmental quality goals would be met? Each of these questions raises a potentially difficult problem. The theoretical and empirical analyses in earlier chapters, however, argue that some of these problems are more apparent than real, and that the others, though serious, are no more severe for exposure charges than for the alternatives, in particular uniform emission standards. Rather than reiterate those arguments, let us compare the operation of standards with exposure charges for a particular class of environmental problems, airborne carcinogens.

Exposure Charges for Airborne Carcinogens

The first step in regulating a carcinogen under Section 112 of the Clean Air Act is to list it as a hazardous air pollutant, as EPA did with benzene in 1977. (By the end of 1982 only six other substances had been listed.) The agency then generates a variety of studies of the health risks and the relative importance of different source categories (e.g., maleic anhydride plants) in contributing to emissions and exposure. For each category that it plans to regulate, EPA must analyze the control options available; that analysis includes engineering studies, cost estimates, and attempts to gauge the impact of alternative regulations on the prices of final products and plant closures.

Based on all of this information, agency officials determine what standard will be proposed, using a variety of ill-defined, often conflicting, criteria. After publication of the proposal, hearings are held and written comments are received from the relevant trade associations, affected firms, environmental groups, and other interested parties. Typically, a wide range of technical and legal objections are made, in response to which EPA may modify or withdraw the proposal. If the standard is promulgated, it is likely to be several years after the whole process began. (Of the seven substances listed under Section 112, final regulations had been promulgated for only four of them by the end of 1982.[1]) By the time of promulgation much of the data on which the proposal was based may have become obsolete, due to changing market conditions, new technologies, or other factors. In the case of benzene almost three years elapsed between the listing and proposal of the first standard. Six years after the listing no final standards had been promulgated; in the meantime several plants had closed, others had installed controls, and still others had stopped using benzene. Following promulgation, the agency must enforce the standard and, often, defend it in court. For many substances most of these steps have to be repeated several times, as separate standards must be set for each type of source.

Now let us consider how exposure charges would work. Listing probably still would be the first step, again followed by a careful review of the health risks and a rough assessment of the relative importance of different source categories. The agency then would have to set the charge rate for that substance.

Setting the Charge

Choosing the level of the charge would be a difficult task, subject to substantial scientific uncertainties and political controversy. The first time an exposure charge was set, special care might be taken. A distinguished panel of scientists, of the type often convened by the National Academy of Sciences, might be asked to review the evidence and make its best estimate of the risk associated with exposure. Empirical estimates of willingness to pay for risk reduction also could provide useful information to decision makers. The analysis in chapter 8, for example, suggests that it would be difficult to justify an exposure charge for benzene in excess of $1 per ppb-person-year. Ultimately, however, although technical information would be an important input, the choice of a particular charge rate would have to rely heavily on the political and ethical judgments of responsible officials, just as the choice of standards now does. Note that only one charge would have to be set for each substance.

Once the first charge had been set, charges for other substances could be set more easily, based on relative potencies, which are subject to less dispute than estimates of absolute risks at low doses. If, for example, a charge of $0.25 per ppb-person-year were set for benzene, and the next substance to be regulated had twice the estimated potency, its charge rate would be $0.50 per ppb-person-year. Consistency in *relative* charges could be achieved without fully resolving the disputes about low-dose extrapolation and the value of risk reduction. The greatest gains in efficiency are likely to come not from getting the absolute levels of the charges "right" but rather from making a more rational and cost-effective allocation of control efforts across substances and sources. The charges should be linked to an index of prices or income to maintain their real incentive effect during times of inflation.

The difficulties involved in setting a charge would be substantial, but they should not be overstated. It is important to remember that the current process of setting standards also is fraught with uncertainty and controversy. EPA does not avoid these difficult scientific, political, and ethical issues by using standards, though it may obscure them. If appropriate standards are to be set, judgments must be made, at least implicitly, about the risks imposed by a substance and about how much we should be willing to spend to reduce those risks. Moreover these judgments must be made repeatedly for each substance, because separate standards must be set for different source categories. With standards EPA also must estimate costs for each source category, a time-consuming and expensive task that yields results of dubious accuracy.

Applying the Charge
After the exposure charge has been set for a substance, EPA would have to determine how to apply it to different types of sources, as exposure itself cannot be measured directly on a continuous basis. For large stationary sources, such as maleic anhydride plants, source-specific dispersion modeling and population data could be used to estimate exposure factors on a source-by-source basis. Such estimates could be made quickly and conveniently with appropriate computer programs to link dispersion models with data bases on meteorological conditions and population. Although dispersion models are of limited accuracy (Miller 1978), the primary determinants of these exposure factors would be population densities, which are measured quite accurately by the Census. The agency would then monitor emissions (as it must now to enforce emission standards); the exposure charge for a facility would be levied on the product of its emissions and exposure factor.

For smaller sources, source-specific exposure factors probably would not be worth estimating. Sources could be grouped, however, by location and other characteristics, with an exposure factor estimated for each category (e.g., service stations in Boston). Often it would be impossible to monitor emissions directly, but indirect measures could be based on more easily monitored parameters. An exposure charge on gasoline, for example, might be collected from wholesalers, based on an estimated exposure factor for a typical service station in the wholesaler's area and the assumption that none of the stations had vapor-recovery devices. Under competitive conditions the charge would be passed on to individual stations (and, ultimately, consumers) in the form of higher gasoline prices. Stations that installed controls could apply for rebates, supplying evidence of the type of control installed and the amount of gasoline pumped.

In many cases the agency would have a range of possibilities for implementing the charge, from very crude methods, to more sophisticated and accurate, but also more complex and expensive, approaches. With automobiles, for example, the simplest approach would be to levy the charge at the time of purchase, basing it on expected lifetime emissions for cars of that type and the exposure factor for the region where the car was sold. A slightly more sophisticated variant would be to levy the charge on an annual basis; manufacturers could indicate the car's emission rate as part of its vehicle identification number. That modification would deal with cars that are moved from the area in which they are originally purchased. Administration of such a charge would be relatively simple if it were collected by state departments of motor vehicles along with existing registration fees or excise taxes. Periodic emission testing might be required (as it soon will be in many states under standards) to provide an incentive to maintain control equipment. If tamper-proof odometers became available at reasonable cost, the charge could also be made a function of the number of miles driven, thus providing encouragement to drive less and the appropriate incentive for high-mileage drivers to purchase low-emission vehicles. Such refinements, however, would be optional and desirable only if the benefits of greater complexity and accuracy exceeded their costs. Indeed, for some source categories exposures might be so low that it would be better to exempt them from the exposure charge altogether, rather than incur even minimal administrative and monitoring costs.

Because of uncertainty about control costs, it would be difficult to predict the precise effects of an exposure charge on environmental quality and risk levels. Some observers may perceive this to be a substantial liability, but in fact it is not. The optimal degree of overall control depends on costs; if they are uncertain, a fixed goal is inappropriate. With expected damages proportional to exposure to airborne carcinogens, the responsiveness of an exposure charge to changes in control costs is highly desirable. For pollutants with sharply nonlinear damage functions, more complex approaches may be appropriate, such as geographically limited permit schemes or differential charges based on ambient concentrations as well as population densities. Even simple uniform exposure charges, however, are likely to be an improvement over uniform emission standards because they direct greater control efforts to high-density areas, which often have more sources and higher ambient concentrations; nonlinear damage functions increase the desirability of controlling such sources.

The revenues raised by charges could be put to a variety of uses. If the federal government retained the charge revenues, it could use them to reduce taxes or for any other purpose. Alternatively, the revenues could be returned to the states or localities where the damages were suffered, thus providing compensation through reduced taxes or expanded public services. By displacing other, distorting taxes—at federal, state, or local levels—charges would provide an additional benefit. It would not be appropriate, in general, to earmark the funds for explicitly environmental purposes; opportunities for public actions to reduce damages from air pollution are almost nil. (With water pollution, public treatment facilities are common, but there is no reason to believe that the optimal amount to spend on such facilities will be equal to the charges collected.)

The system outlined here could be implemented relatively swiftly. Much of the information required already is gathered by EPA, and the rest could be obtained with only modest effort and expense. Conversely, much of the information that the agency now needs to set standards, particularly detailed technical knowledge about control options, no longer would be necessary. Enforcement also need be no more complex than it is for standards. Such a system would be far from perfect. It would not resolve the fundamental uncertainties about how much harm environmental pollutants cause nor the debate about how much we should spend to reduce that harm. It would, however, represent a major improvement over the current system of uniform standards that are inflexible in the face of vast differences across sources in the costs and benefits of control, substantial uncertainty about control costs, and changing conditions.

Notes

Chapter 1

1. For the sources of these estimates, see chapter 9.

2. The economic literature criticizing standards is voluminous. For succinct critiques, see Ruff (1970, 1981). Kneese and Schultze (1975) provide a more extended discussion. Schultze (1977) criticizes the general reliance of U.S. regulation on "command and control" rather than incentives, with numerous environmental examples. Nichols and Zeckhauser (1977, 1981) and Zeckhauser and Nichols (1978) level similar criticisms against OSHA.

3. Mishan (1971) provides a useful survey of "The Post-War Literature on Externalities." See also Baumol and Oates (1975) for a book-length treatment that cites much of the relevant literature. For an important dissent from the Pigouvian tradition, see Coase (1960).

4. Dales (1968) is generally acknowledged as the originator of this approach.

5. Anderson et al. (1977, ch. 6) critically examine many of the political arguments against charges, including the misconception that incentive-based schemes are "licenses to pollute."

6. Drayton (1978, 232–233), for example, labels charges based on damages "the oldest, most classic, most unworkable (even theoretically), and longest and most firmly rejected version of the effluent fee."

Chapter 2

1. Spence and Weitzman (1978, 10), for example, state: "Perhaps the most important single property of pollution damages is that the extra damages of an additional unit of effluent often increase (or at least do not decrease) with the overall level of production. This is sometimes called the 'principle of increasing marginal damages'." As Roberts (1975) and others have pointed out, however, this assumption is invalid in some instances; marginal damages may follow any of many different patterns, depending on specific circumstances.

2. In the literature, "charges," "fees," and "taxes" are used interchangeably. Several officials at EPA have told me, however, that "tax" has undesirable connotations for many individuals —in government, industry, and environmental groups. Thus throughout this study, I refer to "charges." Unfortunately, the obvious literal for the charge rate, c, is more commonly used to represent "cost"; hence, the use of t to stand for the charge (tax) rate.

3. Baumol and Oates (1975, 35), for example, assume that individual utility levels do not depend on the distribution of emissions, only on their sum. As discussed in chapter 6, however, several authors have dealt explicitly with the possibility that marginal damage may vary over time and space.

4. As noted in chapter 1, Dales (1968) is generally credited with originating the idea of marketable permits. In the literature, "marketable rights," "marketable licenses," and a variety of similar terms also are often used.

5. The study was conducted in the mid-1960s for a special interdepartmental committee on water quality control (Kneese and Bower 1968, 158). The figures given are those cited by Kneese and Bower (1968, 162).

6. Other authors also have used specific functional forms in their analyses. Weitzman (1974), for example, in essence assumes that control costs are a quadratic function of the degree of emission control. Morgan et al. (1978) assume that control costs for sulfur control are given by $C(r) = a \log[1/(1 - r)]$, where r is the degree of control and a is distributed normally across plants. If we let $x = 1 - r$ be the level of emissions not controlled, then $C(x) = -a \log x$. The particular cost function used in this study approaches Morgan et al.'s logarithmic function as α approaches 0.

7. This formulation ignores the fact that as permitted emission levels approach zero and control costs grow without bound, firms have the option of shutting down the source, in which case the marginal cost of further control becomes zero. See Starrett and Zeckhauser (1974) and Baumol and Bradford (1972) for discussions of how shutdown possibilities can introduce nonconvexities into both costs and damages.

8. In particular, the ratio of the social cost of the charge to that of the standard, S^*/S_s^*, is the ratio of the linear case raised to the power $\beta(\alpha + 1)/(\alpha + \beta)$, which is greater than unity if $\alpha > 0$ and $\beta > 1$ (e.g., if $\alpha = 1$ and $\beta = 2$, $\beta(\alpha + 1)/(\alpha + \beta) = 4/3$).

9. See virtually any mathematical statistics text (e.g., Hogg and Craig 1970, 105) for a derivation of the moment-generating function for the normal distribution.

10. With $\sigma_a^2 = 2$, firms in the 10th percentile have costs that are 1/6 those of the median firm, whereas firms in the 90th percentile have costs that are 6 times higher. Average cost is 2.7 times higher than the median firm's. With $\sigma_a^2 = 1$, average cost is 1.6 times higher than the median firm's.

11. If we employed the more usual procedure and measured the benefit of switching from a standard to a charge as the savings in control costs for achieving a given level of emissions, the savings is given by $\{1 - [E(a^\varepsilon)]^{1/\varepsilon}/E(a)\}$. With the log-normal distribution the savings is $\Delta S = \{1 - \exp[-0.5(1 - \varepsilon)\sigma^2]\}$, as compared to equation (2.22). For $\sigma_a^2 = 2$, the savings are then 39 percent for $\alpha = 1$ and 28 percent for $\alpha = 0.5$.

Chapter 3

1. This principle has been recognized in a variety of contexts, on occasion by politicians as well as economists. David Lloyd George, for example, reportedly noted in connection with workmen's compensation that "the cost of the product should bear the blood of the working-man" (Nichols and Zeckhauser 1981).

2. For efficiency it is important to pay the subsidy to potential entrants as well as to firms in existence; otherwise the subsidy will provide an inappropriate incentive for existing firms to stay in business and for new firms to enter. Moreover, as Polinsky (1979) shows, a lump-sum subsidy that is contingent on operation also leads firms to choose an inefficiently small scale of operation; the lump-sum payment reduces fixed costs and hence both lowers average total cost and shifts its minimum to a lower level of output.

3. This is important not only to ensure that each firm does not pay more than its damages but also that it faces the right marginal incentive. A single firm facing an upward sloping "supply curve" (the marginal damage curve) is a monopsonist and will "buy" too little (will emit too little); it will view the supply curve as an average price curve, with the marginal price of emissions at a higher level reflecting the fact that, when it emits more, it must pay a higher price for inframarginal emissions.

4. For a basic survey of these issues, see Musgrave and Musgrave (1976, chs. 21, 22).

5. Note than an emission charge, unlike most taxes, constitutes the entire price, not just some

fraction. Thus the conditions under which a price change will increase charge collections are simply those that apply to a firm seeking to alter its price to increase its revenues.

6. If demand is given by $Q = bP^{-\varepsilon}$, where ε is the (absolute value of the) elasticity of demand, changing P by a factor of k changes Q by a factor of $k^{-\varepsilon}$: $2^{-0.5} = 0.707$ (29.3 percent decrease) and $0.8^{-2} = 1.563$ (56.3 percent increase).

7. Terkla (1979, ch. 4) estimates that a charge of $0.25 per pound of sulfur dioxide emitted from major stationary sources would generate approximately $1 billion to $2.5 billion per year (in 1978 dollars) in tax-displacement efficiency gains, depending on the particular taxes reduced with the charge revenues.

8. EPA, for example, has not yet decided whether it will auction or allocate free of charge marketable permits if it uses them to control chloroflourocarbons (Rabin 1981). EPA's existing "offset" program for "prevention of significant deterioration" of air quality in essence grants a type of marketable permit to existing emitters, free of charge (Repetto 1983).

9. As with many equity-based arguments this one is clouded by uncertainty about the ultimate burden of different forms of taxation. Although firms would pay an emission charge, it seems unlikely that their owners would bear most of the burden in the form of reduced profits. Similarly, how much of the burden of the corporate income tax falls on owners of capital is an open question. One well-known, but controversial, study (Krzyzaniak and Musgrave 1963) suggests that none of the burden is borne by capital.

Note that it would not be desirable to use the revenues to provide *differential* tax relief to pollution-intensive industries that had unusually high pollution-charge payments, as that would have the effect of lowering costs and prices in such industries, thus partially undoing one of the beneficial impacts of pollution charges.

Chapter 4

1. See Page (1978) for a discussion of the asymmetries in the costs of "false positives" and "false negatives" in regulating toxic chemicals. He argues that current regulatory practice places too much emphasis on avoiding false positives (incorrectly identifying substances as hazardous). I disagree with his interpretation of current behavior by regulators, but I agree with his conclusion that regulators should use the expected risk in making decisions, thus taking account of both the magnitudes of the consequences and the probabilities.

2. Slesin and Ferreira (1976), for example, argue that "empirical evidence, based on frequencies of multiple-death accidents in the United States (1956–1970), suggests that the social impact of major accidents varies with the cube of the number of lives lost." That is, an accident killing 1,000 people is 1 million times worse than 1,000 separate accidents killing 1 person each. Wilson (1974) assumes, "as a guess, that a risk involving N people simultaneously is N^2 (not N) times as important as an accident involving one person." Under Wilson's rule of thumb, an accident involving 1,000 deaths is 1,000 times worse than 1,000 accidents each involving 1 death.

3. Schelling (1968), for example, suggests that one of the major "costs" of premature death is the grief caused the family and close friends of the victim. Thus catastrophes that concentrate deaths in one area may be preferred to an equal number of deaths scattered among different clusters of relatives and friends. A simple thought experiment suggests the merit of this argument. Suppose you have a spouse and two children. You have a choice in making a hazardous journey: all of you can travel together, in which case there is a 0.01 chance you will all die, or you can travel separately, each facing an independent risk of 0.01. Both options yield the same number of deaths, $4(0.01) = 0.04$, but I suspect most readers would prefer the first.

4. One caveat should be noted. Most of the uncertainty about low-dose carcinogenic risks is based on uncertainty about the correct model to use for low-dose extrapolation, as discussed in more detail in chapter 8. Thus the risks posed by different chemicals may be highly correlated, and even in the aggregate there may be substantial variation in the number of deaths due to chemical carcinogens. Nonetheless, that variation is likely to be small relative to all deaths, or even to all cancer deaths.

5. This possibility is a little farfetched, as the data in chapter 8 make apparent, though not far from what some advocates would have us believe. At an informal hearing on EPA's listing of benzene as a hazardous air pollutant that I attended in early 1978, a representative of a well-known environmental group suggested that EPA should take the total number of deaths from leukemia and subtract the deaths attributable to other specific causes (this number would be very small). The environmentalist suggested that EPA then use the residual as an upper-bound estimate of the leukemia deaths due to benzene.

6. The final CAG report (Albert et al. 1979) did not include this qualifying statement; indeed, the tone of the final report suggests that the estimate is a best estimate, and not a conservative one. The change between the two versions of the report is a typical illustration of how assumptions and fundamental uncertainties tend to get lost along the way.

7. In traditional academic terms, the conservative case for regulation would be made by showing that, even if we use assumptions that bias the risk estimate downward, the benefits of regulation still exceed the costs.

8. Nichols (1975), for example, presents a case involving an EPA study of the health effects associated with long-lived radionuclides from nuclear power plants. The rankings of the hazards posed by various radionuclides differ greatly depending on whether one uses the figures based on "anticipated minimum performance by industry" or the "maximum plausible" estimates advocated by EPA for public health planning.

9. If the distributions are negatively correlated, the estimate will be more than eight times higher then the expected value. If they are positively correlated, the estimate will be less than eight times higher. In either case the expected risk will not be the product of the expected value of the three parameter values. In fact, in most instances, it seems likely that the distributions will be independent. Continuing with the illustration in the text, it seems unlikely that uncertainties about emissions, exposure, and the dose-response relationship will be interdependent.

10. This approach encounters several difficult obstacles, including: (1) how to elicit probability estimates from experts, (2) how to combine estimates from different experts with epidemiological and experimental data, and (3) how to endow the process with sufficient credibility to withstand political and judicial scrutiny. Raiffa and Zeckhauser (1981) offer a useful, but not conclusive, discussion of these and other related issues.

11. The basic framework that follows draws heavily on Weitzman's (1974) analysis of a central planner's choice between sending price or quantity signals to firms under conditions of uncertainty. Weitzman's analysis of a single firm can be extended to multiple firms if the "quantity" instrument is marketable permits. Rose-Ackerman (1974) offers a similar graphical analysis.

12. As reported in chapter 9, for example, the Chemical Manufacturers' Association estimates much higher costs than EPA for controlling benzene emissions from maleic anhydride plants. On the other side, Mark Green (1979), director of Public Citizen's Congress Watch, argues that, because industry controls the data on which regulatory cost estimates are based, the estimates are too high.

13. Note, again, that an emission standard does not guarantee that a particular level of total

emissions will be achieved, as it controls only the emission rate and not the number of firms or production levels.

14. Weitzman (1974) reaches quite a different conclusion, because in his formulation the central authority is uncertain about the absolute level of marginal damage but not about its rate of change. In the case of absolute thresholds, for example, Weitzman's planner is uncertain as to how much damage will be caused if the threshold is exceeded but knows with certainty the threshold level of emissions. This uncertainty has no effect on the slope of the expected marginal damage function. This assumption, however, seems highly unrealistic in relation to most environmental problems, where the location of the threshold (indeed, its very existence) usually is hotly debated and highly uncertain.

15. The cost estimates for controlling benzene from maleic anhydride plants, discussed in chapter 9, provide an excellent example. One division of EPA had a consulting firm prepare a crude cost estimate (Energy and Environmental Analysis, Inc. 1978). At the same time, another division had another contractor prepare a more detailed study (Lawson 1978) that was later updated for use in the environmental impact statement (U.S. EPA 1980). The agency claimed an accuracy of ±30 percent, but, as discussed in chapter 9, testimony by the Chemical Manufacturers' Association suggests the estimates may be in error by a far larger factor.

16. Drayton (1978) argues that adjustment is less likely with charges, because legislators will insist on setting the rates. In contrast, he claims, "administrators can adjust their regulations more or less as needed." His view of standards seems excessively rosy given the capital-intensive nature of pollution control; small changes in standards for existing sources may require installation of entirely new equipment.

17. If marginal damage is invariant with respect to emissions, a 10 percent increse in marginal damage (e.g., due to an increase in population) will cause a 10 percent decrease in the optimal level of emissions only if the elasticity of demand for the right to emit with respect to the price of emissions is unity. If demand is inelastic, emissions should fall less than 10 percent. If it is elastic, emissions should decrease by more than 10 percent.

18. Kneese and Bower (1968, 139) note: "Under an effluent standards system, the waste discharger has no incentive to do more than meet the standard."

19. If the standards change in response to innovation (e.g., EPA's best available technology approach), *suppliers* of emission control equipment may have a very strong incentive to innovate under standards; a supplier that develops and patents a new technique that can achieve hitherto impossible levels of control—and then can induce EPA to impose that level of control through a new standard—can reap large profits or rents. See Repetto (1980) for a discussion of the role of suppliers in pollution abatement innovation.

Chapter 5

1. Some readers may object that this characterization of the literature is unfair, that many authors have recognized that the damage caused by emissions varies across time and space. Several studies that deal with the issue formally are cited in chapter 6, and many other studies at least mention it in passing. It seems fair to say, however, that this issue rarely has been explored in depth. Anderson et al. (1977), for example, devote less than a page to it in a book-length treatment of economic incentives. Ruff (1970), in a well-known, non-technical article deals with the issue in a single sentence: "In principle, the prices should vary with geographical location, season of the year, direction of the wind, and even day of the week, although the cost of too many variations may preclude such fine distinctions."

2. Under OSHA's concentration standards a firm receives no credit for reducing the number of workers who are exposed to a hazardous substance, say, through greater automation. Note that a charge-based approach, where the fee is based on the product of the number of workers and the average exposure level, provides an incentive to reduce the number of workers exposed.

3. The damages caused by emissions also may vary depending on a variety of other conditions. With water pollution, for example, the amount of water flowing in a river, which varies seasonally, is often important (Roberts 1975). With aircraft noise the time of day is critical; noise is far more annoying at night than during the day and early evening (Harrison 1983).

4. In certain cases, however, input-based charges or permits may be relatively efficient. With chloroflourocarbons, for example, all of the substance eventually is emitted, and the level of damages appears to be independent of the locations of the emissions (Rabin 1981). Thus a simple charge on the production of chloroflourocarbons, or a marketable permit system imposed on manufacturers, is likely to be quite efficient. (If we discount future health effects, however, a uniform charge or undifferentiated permit system is not efficient; the charge should be higher for those uses where the release is immediate and lower for those where it is delayed until many years after the date of manufacture.)

5. The calculation is straightforward. The probability of zero injuries is simply $0.99^{100} = 0.366$. The probability of exactly one injury is $\binom{100}{1}(0.99)^{99}(0.01) = 100(0.99)^{99}(0.01) = 0.370$. Thus the probability that two or more will be injured is $1 - (0.366 + 0.370) = 0.264$.

6. As Settle and Weisbrod (1977) note, "It is always possible to translate a tax into a standard —or, alternatively, a standard into a tax—by equating the tax function and the penalty function of the standard."

7. Crudely put, the theory of the second-best states that, if some optimality conditions cannot be satisfied, in general it will not be efficient to satisfy the others (Lipsey and Lancaster 1956–57). In this context the theory of the second best suggests, for example, that if we are unable to use the best instrument, the second-best strategy may not involve the same target as the optimal, but unattainable, strategy.

Chapter 6

1. More specifically, Atkinson and Lewis (1974) estimate that the cost of achieving the primary standard with an ambient least-cost strategy would be about one-half the cost under an emission least-cost strategy, which in turn would cost about one-sixth as much as a uniform roll back strategy, as reported in chapter 2.

2. Kneese and Bower (1968, 162) report that, to achieve a dissolved oxygen objective of 3 to 4 ppm would have cost (in millions of dollars per year) 7.0 under the least-cost strategy, 8.6 with a three-zone charge scheme, 12.0 with a single charge, and 20.0 with uniform treatment.

3. Radiation exposure limits are stated in terms of maximum exposure per year. Some firms took advantage of this fact by hiring transient workers for very short periods of time (some for only half a day) to work with radioactive materials. Under such conditions some workers received the maximum allowable annual dose in a few minutes (Gillette 1974). The firms stayed in compliance with the regulations, with no individual worker being overexposed, but total exposure levels were undoubtedly higher than they would have been if there had not been such rapid turnover in the work force.

4. The loss in efficiency from restricting the number of classes for the conditional standard

does not appear to be very great. Using the model developed in this chapter, with $\sigma_a^2 = 2$, $\rho = 0$, and $\alpha = 1$ or 0.5, moving from one class (a uniform standard) to two classes provides 66 percent of the efficiency gains of moving from one class to an infinite number. Three classes yield more than 82 percent of the potential gain, and with four classes we obtain over 89 percent of the maximum improvement possible with the conditional standard. Thus it appears that refining the exposure classes yields rapidly diminishing returns.

5. Some readers may worry about the firm that locates in a lightly populated area, only to find that the population grows over time. This possibility raises two issues: (1) Is it fair to increase the charge rate as the population grows? (2) Should measures be taken to discourage people from moving to the area? The answer to the second question is almost certainly no. As Baumol and Oates (1975) show, the externality itself provides an approiate incentive for people not to move in; no additional incentive is necessary or desirable. The answer to the first question depends on one's personal values. I believe that it is not unfair to raise the charge as the population expands. Consider the following analogy. A firm locates in an economically depressed town to take advantage of low wages and land rents. The town expands, new businesses enter, and competition drives up wages and land rents. Is it unfair for the company to have to pay these higher prices?

Chapter 7

1. It is important to note that, if the two types of cost coefficients are not misestimated by the same factor, the performance standard is less affected than the specification standards and, in some cases, the specification charges. If, for example, a is overestimated by factor of two and b is underestimated by factor of two, the performance standard will not be set at the wrong level, but the specification charges and standards will be set incorrectly.

2. The marginal damage caused by x will be λy, hence the charge collected on x will be $x(\lambda y) = \lambda x y$. Similarly, the charge collected on y will be $y(\lambda x) = \lambda x y$. Thus the total collected will be $2\lambda x y$, double the actual damage.

3. This fact is not apparent in table 7.2 because the efficiencies have been normalized so that the efficiency of the performance standard is always 0, regardless of the value of ρ. It can be shown, however, that the more negative ρ is, the lower the absolute costs under the two performance-based strategies.

Chapter 8

1. OSHA issued an emergency temporary standard in May 1977 and also proposed a new permanent rule, both of which lowered the exposure limit from 10 to 1 ppm. The new permanent standard was promulgated in February 1978 (43 *Fed. Reg.* 5918 1978) but struck down later that year by the Fifth Circuit Court of Appeals on the grounds that OSHA had failed to show that the standard yielded significant benefits (581 F.2d 493 1978). The Supreme Court upheld the lower court's decision in 1980 (100 S.Ct. 2844 1980).

2. Gasoline averaged an estimated 1.3 percent benzene in 1977 (Turner et al. 1978, 1-5). Annual U.S. gasoline consumption is on the order of 100 billion gallons (Mara and Lee 1978, 112), so the total amount of benzene in gasoline in 1977 was about 1.3 billion gallons, as compared to 1979 production of about 1.7 billion gallons.

3. See Nichols (1981, ch. 8) for a discussion of these problems, which ranged from questionable aggregation of the exposure estimates for automobiles to errors in converting data in square miles to square kilometers.

4. PEDCo estimated that benzene emissions from maleic anhydride plants were 34.8 million pounds per year, or about 15.8 million kilograms. PEDCo rated the accuracy of that estimate higher than for any other source category. The draft Environmental Impact Statement for maleic anhydride plants, which was based on more complete information, estimated the total at only 5.8 million kilograms at full-capacity operation; PEDCo overestimated the uncontrolled emission rate and did not take account of the fact that several plants already had controls. In general, the emission and exposure estimates discussed in this section appear to have been too high. See Nichols (1981, ch. 8) for more details.

5. Jandl (1977), for example, argues that benzene-induced leukemia is always preceded by aplastic anemia and that it occurs only at doses exceeding 100 ppm. For a balanced, non-technical discussion of the controversy about carcinogens and thresholds, see Maugh (1978).

6. The number of animals required in laboratory experiments to detect low-level risks is simply too large (Schneiderman, Mantel, and Brown 1975). A variety of problems limit the sensitivity of epidemiological studies (Calkins et al. 1979).

7. The literature on these models is voluminous. For a useful review, see Food Safety Council (1978, ch. 11).

8. More specifically, the risk at dose D is $1 - \exp(-\gamma D)$, where γ is estimated empirically.

9. The risk at dose D is $\Phi(\alpha + \beta \log D)$, where $\Phi(\cdot)$ is the cumulative unit normal distribution and α and β are empirical constants.

10. Under the logit model, the risk is $[1 - \exp(-\alpha - \beta \log D)]^{-1}$. Under the k-hit model, it is

$$\int_0^D \left[\frac{x^{k-1}e^{-x}}{\Gamma(k)} \right] dx,$$

where $\Gamma(\cdot)$ is the gamma function.

11. The CAG estimated confidence limits around its estimate of total fatalities (Albert et al. 1979, 23). The estimate in the text merely converts the CAG figures to risk per unit of exposure.

12. Readers versed in applied welfare economics will recognize the two measures as compensating variation and equivalent variation, which in general will differ because of income effects. If the income effects are neglible, as they will be with small risks, the two measures yield virtually identical results.

13. For a useful and balanced discussion of discounting morbidity and mortality measures, see Raiffa, Schwartz, and Weinstein (1977, sec. 4.4).

14. Most cost-effectiveness analyses of life-saving programs ignore the time element, thus implicitly applying a zero discount rate. In their analysis of carcinogens in drinking water, Page, Harris, and Bruser (1979) argue explicitly in favor of comparing steady-state costs and benefits (which yields essentially the same result as a zero discount rate) on the grounds of intergenerational equity.

15. The calculations are straightforward:

$$\sum_{t=0}^{34} \frac{58,164}{1.05^t} = 1 \times 10^6,$$

and

$$\sum_{t=15}^{34} \frac{58,164}{1.05^t} = 366,100.$$

Chapter 9

1. One of the eight benzene-fed plants (Monsanto) uses *n*-butane for approximately 20 percent of its capacity. Another plant uses *n*-butane exclusively, and one recovers maleic anhydride in the process of manufacturing phthalic anhydride; neither emits any benzene.

2. These are nominal control levels; EPA assumes that with the addition of controls, conversion rates will fall from an average of 94.5 to 90 percent, thus increasing the uncontrolled emission rate. As a result 97 percent control would yield a net reduction of only 94.6 percent, and 99 percent cuts emissions by only 98.2 percent.

3. The basic technical data are in the contractor's report (Lawson 1978). EPA's estimates in the draft environmental impact statement adjust the costs to reflect more recent prices. The original estimates indicated that incineration was less expensive, but a sharp rise in the price of benzene (and hence an increase in the credit for benzene recovered with adsorption) reversed the ranking. See U.S. EPA (1980, table 5-4) for a comparison of the old and new parameter values.

4. For carbon adsorption at 97 percent, net annualized costs are about 4 percent higher at 56 percent capacity utilization than at full capacity. With thermal incineration costs fall with capacity utilization, though not in proportion (U.S. EPA 1980, 5-21 to 5-24).

5. These exposure factors were estimated from data in U.S. EPA (1980 app. E). The dispersion modeling used meteorological data from Pittsburgh, which give higher than average concentrations, as noted in chapter 4.

6. The cost-effectiveness ratio of shifting from 90 to 99 percent control is only $10.80 per ppb-person-year, however, so there is no value of V for which 97 percent is optimal.

7. In all cases but Tenneco, 97 percent control is not in the rankings because its cost-effectiveness ratio is higher than that for 99 percent.

8. All the testimony by CMA witnesses cited here was presented orally at hearings held on August 21, 1980 in Washington, D.C. The figures are drawn from the written version of that testimony, "CMA Presentation at the EPA Hearings on the Proposed NESHAP for Benzene Emissions from Maleic Anhydride Plants," August 21, 1980.

9. Galluzzo and Glassman (1980) also noted that Koppers and Reichold (N.J.) had closed, but both of those plants already met the proposed standard, so no costs or benefits were included for them in our original analyses, with the exception of Reichold (N.J.) for the uniform 99 percent standard.

10. With an annual capacity of 10.7×10^6 kg maleic anhydride per year, the Pfizer plant is smaller than any of the others. Control costs for 97 percent were estimated at $199,000 using a linear extrapolation from the two smallest plants analyzed by EPA—Tenneco and Reichold (Ill.). The estimated reduction in emissions at 97 percent control is 678,000 kg of benzene per year, based on EPA's assumptions about uncontrolled emissions, the conversion rate, and so on. The estimated reduction in exposure is 11,119 ppb-person-years, based on concentrations from EPA's model plant (U.S. EPA 1980, E-12) scaled down to the smaller plant size and on population estimates presented by Galluzzo and Glassman (1980).

11. The CMA estimates should be viewed with some skepticism, as they come from parties with an incentive to overstate the costs. At the public hearings, however, CMA witnesses presented several pieces of corroborating evidence that suggest, at the very least, that the EPA estimates are too low.

12. This new information was obtained in a telephone conversation with Rick Colyer of EPA's Office of Air Quality Planning and Standards, Durham, N.C.

Chapter 10

1. The seven substances and the years in which they were listed are beryllium (1971), asbestos (1971), mercury (1971), vinyl chloride (1975), benzene (1977), radionuclides (1979), and inorganic arsenic (1980). Final standards for the first three were promulgated in 1973. The vinyl chloride standard was promulgated in 1976 (Haigh 1982b). As noted in chapter 8, four standards for benzene source categories were proposed in 1980 and 1981. By the end of 1982 none of those had been promulgated, nor had any additional standards been proposed for any of the other listed substances.

References

Acton, J. "Evaluating Public Programs to Save Lives." Santa Monica, Calif.: Rand Corporation, No. R-950-RC, January 1973.

Aksoy, M., S. Erdem, and G. Dincol. "Leukemia in Shoe Workers Exposed Chronically to Benzene," *Blood* 44 (1974): 837–841.

Aksoy, M., S. Erdem, and G. Dincol. "Types of Leukemia in Chronic Benzene Poisoning: A Study in Thirty-Four Patients." *Acta Haematalo* 55 (1976): 65–72.

Aksoy, M. Testimony before Occupational Safety and Health Administration. Washington, D.C., July 13, 1977.

Albert, R., E. Anderson, I. Dubin, R. Hill, R. McGaughy, L. Mishra, R. Pertel, W. Richardson, and T. Thorslund. "Carcinogen Assessment Group's Preliminary Report on Population Risk to Ambient Benzene Exposure." Unpublished paper, U.S. Environmental Protection Agency, 1977.

Albert, R., E. Anderson, C. Hiremath, R. McGaughy, S. Miller, R. Pertel, W. Richardson, D. Singh, T. Thorslund, and A. Zahner. "Carcinogen Assessment Group's Final Report on Population Risk to Ambient Benzene Exposures." Unpublished paper, U.S. Environmental Protection Agency, January 10, 1979.

Anderson, F., A. Kneese, P. Reed, S. Taylor, and R. Stevenson. *Environmental Improvement through Economic Incentives*. Baltimore: The Johns Hopkins University Press, 1977.

Atkinson, S. E., and D. II. Lewis. "A Cost-Effectiveness Analysis of Alternative Air Quality Control Strategies." *Journal of Environmental Economics and Management* 1 (1974): 237–250.

Bailey, M. J. *Reducing Risks to Life*. Washington, D.C.: American Enterprise Institute, 1980.

Baumol, W., and D. Bradford. "Detrimental Externalities and Non-Convexity of the Production Set." *Economica* 39 (1972): 160–176.

Baumol, W. J., and W. E. Oates. "The Use of Standards and Prices for Protection of the Environment." In P. Bohm and A. V. Kneese (eds.), *The Economics of Environment*. New York: St. Martin, 1971, pp. 53–65.

Baumol, W. J., and W. E. Oates. *The Theory of Environmental Policy—Externalities, Public Outlays, and the Quality of Life*. Englewood Cliffs, N.J.: Prentice-Hall, 1975.

Blomquist, G. "Valuation of Life: Implications of Automobile Seat Belt Use." Ph.D. dissertation. University of Chicago, 1977.

Browning, E. "The Marginal Cost of Public Funds." *Journal of Political Economy* 84 (1976): 283–298.

Calkins, D., R. Dixon, C. Gerber, D. Zarin, and G. Omenn. "Identification, Characterization, and Control of Potential Human Carcinogens: A Framework for Federal Decision-Making." Staff paper, Office of Science and Technology Policy, Washington, D.C., February 1, 1979.

Chemical Manufacturers' Association. "Cost Considerations." In *CMA Presentation at the EPA Hearing of the Proposed NESHAP for Benzene Emissions from Maleic Anhydride Plants*. Washington, D.C., August 21, 1980.

Clark, T. B. "New Approaches to Regulatory Reform—Letting the Market Do the Job." *National Journal* (August 11, 1979): 1316–1322.

Coase, R. H. "The Problem of Social Cost." *Journal of Law and Economics* 3 (1960): 1–44.

Dales, J. H. *Pollution, Property and Prices.* Toronto: University of Toronto Press, 1968.

Dillingham, A. E. "The Injury Risk Structure of Occupations and Wages." Ph.D. dissertation. Cornell University, 1979.

Drayton, W. "Comment." In A. F. Friedlaender (ed.), *Approaches to Controlling Air Pollution.* Cambridge: The MIT Press, 1978, pp. 231–239.

Energy and Environmental Analysis, Inc. "Estimated Cost of Benzene Control for Selected Stationary Sources." Report submitted to Strategies and Air Pollutant Standards Division, U.S. Environmental Protection Agency, February 27, 1978.

Feldstein, M. "The Welfare Cost of Capital Income Taxation." *Journal of Political Economy* 86 (1978): S29–S53.

Food Safety Council. *A System for Food Safety Assessment.* Final report of the Scientific Committee, Columbia, Md., June 1978.

Galluzzo, N. B., and D. Glassman. "Recalculated Benzene Emissions, Population Exposures, and Risk." In *CMA Presentation at the EPA Hearing on the Proposed NESHAP for Benzene Emissions from Maleic Anhydride Plants.* Chemical Manufacturers' Association, Washington, D.C., August 21, 1980.

Gillette, R. "'Transient' Nuclear Workers: A Special Case for Standards," *Science* 186 (1974): 125–129.

Green, M. "The Fake Case Against Regulation." *Washington Post* (January 21, 1979): C1.

Haigh, J. "EPA and the Regulation of Coke Oven Emissions: A Cost-Benefit Analysis." Unpublished paper, Harvard University, April, 1982a.

Haigh, J. "History of Section 112 of the Clean Air Act Amendments of 1970." Draft paper, Harvard University, September, 1982b.

Harrison, D. *Who Pays for Clean Air.* Cambridge: Ballinger, 1975.

Harrison, D. "The Regulation of Aircraft Noise." In T. C. Schelling (ed.), *Incentives for Environmental Protection.* Cambridge: The MIT Press, 1983, pp. 41–143.

Hogg, R. V., and A. T. Craig. *Introduction to Mathematical Statistics,* 3rd ed. New York: Macmillan, 1970.

Infante, P., R. Rinsky, J. Wagoner, and R. Young. "Leukemia in Benzene Workers." *Lancet* 2 (July 1977): 76–78.

Jandl, J. H. "A Critique of EPA's Assessment of Health Effects Associated with Atmospheric Exposure to Benzene." Submitted by the American Petroleum Institute to U.S. Environmental Protection Agency, December 1977.

Kneese, A. V., and B. T. Bower. *Managing Water Quality: Economics, Technology, Institutions.* Baltimore: The Johns Hopkins University Press, 1968.

Kneese, A. V., and C. L. Schultze. *Pollution, Prices and Public Policy.* Washington, D.C.: Brookings Institution, 1975.

Krzyzaniak, M., and R. A. Musgrave. *The Shifting of the Corporation Income Tax.* Baltimore: The Johns Hopkins University Press, 1963.

Lamm, S. H. "Oral Presentation to the EPA for the American Petroleum Institute." Submitted to U.S. Environmental Protection Agency, August 21, 1980.

Lawson, J. F. "Maleic Anhydride—Product Report." Report prepared by Hydroscience, Inc., for U.S. Environmental Protection Agency, Contract no. 68-02-2577, 1978.

Leone, R. A., and J. E. Jackson. "The Political Economy of Federal Regulatory Activity." Working paper, Graduate School of Business, Harvard University, March 1978.

Lipsey, R. G., and K. Lancaster. "The General Theory of Second Best." *Review of Economic Studies* 24 (1956–57): 11–32.

Luken, R., and S. Miller. "Regulating Benzene: A Case Study." Unpublished paper, U.S. Environmental Protection Agency, September 1979.

Mantel, N., and W. R. Bryan. "'Safety' Testing of Carcinogenic Agents." *Journal of the National Cancer Institute* 27 (August 1961): 455–470.

Mara, S. J., and S. S. Lee. *Human Exposures to Atmospheric Benzene*. Report prepared by Stanford Research Institute for U.S. Environmental Protection Agency, October 1977.

Mara, S. J., and S. S. Lee. *Assessment of Human Exposure to Atmospheric Benzene*. Report prepared by SRI International for U.S. Environmental Protection Agency, May 1978.

Maugh, T. H. "Chemical Carcinogens: How Dangerous are Low Doses." *Science* 202 (October 1978): 37–41.

Miller, C. *Exposure Assessment Modelling: A State-of-the-Art Review*. Report prepared for U.S. Environmental Protection Agency, No. EPA-600/3-78-065, 1978.

Mishan, E. J. "The Postwar Literature on Externalities: An Interpretative Essay." *Journal of Economic Literature* 9 (March 1971): 1–28.

Montgomery, W. C. "Markets in Licenses and Efficient Pollution Control Programs." *Journal of Economic Theory* 5 (1972): 395–418.

Morgan, G., S. Morris, W. Rish, and A. Meier. "Sulfur Control in Coal Fired Power Plants: A Probabilistic Approach to Policy Analysis." Unpublished paper, Carnegie-Mellon University, Pittsburgh, June 1978.

Musgrave, R. A., and P. B. Musgrave. *Public Finance in Theory and Practice*, 2nd ed. New York: McGraw-Hill, 1976.

National Academy of Sciences. "Health Effects of Benzene: A Review." Report on Committee of Toxicology, Washington, D.C., June 1976.

Nichols, A. L. "The Use of Benefit-Cost Analysis in Setting Low-Level Radiation Standards." Unpublished paper, Harvard University, June 1975.

Nichols, A. L. "Choosing Regulatory Targets and Instruments, with Applications to Benzene." Ph.D. dissertation, Harvard University, 1981.

Nichols, A. L., and R. Zeckhauser. "Government Comes to the Workplace: An Assessment of OSHA." *Public Interest* 49 (Fall 1977): 39–69.

Nichols, A. L., and R. Zeckhauser. "OSHA after a Decade: A Time for Reason." In L. Weiss and M. Klass (eds.), *Case Studies in Regulation*, pp. 202–234. Boston: Little, Brown, 1981.

Ott, M., J. Townsend, W. Fishback, and R. Langer. "Mortality Among Individuals Occupationally Exposed to Benzene." Exhibit 154, OSHA Benzene Hearings, July 19–August 10, 1977.

Page, T. "A Generic View of Toxic Chemicals and Similar Risks." Social Science Working Paper No. 198, California Institute of Technology, Pasadena, January 1978.

Page T., R. Harris, and J. Bruser. "Removal of Carcinogens from Drinking Water: A Cost-Benefit Analysis." Social Science Working Paper No. 230, California Institute of Technology, Pasadena, January 1979.

PEDCo Environmental, Inc. *Atmospheric Benzene Emissions.* Report submitted to U.S. Environmental Protection Agency, Research Triangle Park, N.C., EPA-450/3-77-029, October 1977.

Pigou, A. C. *The Economics of Welfare,* 4th ed. London: Macmillan, 1932.

Plott, C. "Externalities and Corrective Taxes." *Economica* 33 (1966): 84–87.

Polinsky, A. M. "Notes on the Symmetry of Taxes and Subsidies in Pollution Control." *Canadian Journal of Economics* 12 (February 1979): 75–83.

Pratt, J., and R. Zeckhauser. "Incentive-Based Decentralization: Expected Externality Payments Induce Efficient Behavior in Groups." Unpublished paper, Harvard University, February 1981.

Rabin, R. L. "Ozone Depletion Revisited." *Regulation* (March–April 1981): 32–38.

Raiffa, R. L. *Decision Analysis.* Reading, Mass.: Addison-Wesley, 1968.

Raiffa, H., W. B. Schwartz, and M. C. Weinstein. "Evaluating Health Effects of Societal Decisions and Programs." in *Decision Making in the Environmental Protection Agency,* vol. 2B. Washington, D.C.: National Academy of Sciences, 1977.

Raiffa, H., and R. Zeckhauser. "Reporting of Uncertainties in Risk Analysis." Unpublished paper, Harvard University, April 1981.

Redmond, C., B. Strobino, and R. Cypress. "Cancer Experience Among Coke By-product Workers." *Annals of the New York Academy of Sciences* 271 (1976): 102–115.

Regulatory Analysis Review Group. "U.S. Environmental Protection Agency's Proposed Policy and Procedures for Identifying, Assessing and Regulating Airborne Substances Posing a Risk of Cancer." Report Submitted by Council on Wage and Price Stability, February 21, 1980.

Repetto, R. "The Influence of Standards, Efficient Charges and Other Regulatory Approaches on Innovation in Abatement Technology." Unpublished paper, Harvard University, September 1980.

Repetto, R. "Air Quality Under the Clean Air Act" In T. C. Schelling (ed.), *Incentives for Environmental Protection.* Cambridge: The MIT Press, 1983, pp. 223–290.

Roberts, M. J. "Environmental Protection: The Complexities of Real Policy Choice." In I. K. Fox and N. A. Swainson (eds), *Water Quality Management: The Design of Institutions.* University of British Columbia Press, 1975, pp. 157–235.

Rose-Ackerman, S. "Effluent Charges: A Critique." *Canadian Journal of Economics* 6 (1973): 512–518.

Ruff, L. "The Economic Common Sense of Pollution." *The Public Interest* (Spring 1970): 69–85.

Ruff, L. "Federal Environmental Regulation." In L. Weiss and M. Klass (eds.), *Case Studies in Regulation.* Boston: Little, Brown, 1981, pp. 235–261.

Schelling, T. C. "The Life You Save May Be Your Own." In S. Chase (ed.), *Problems in Public Expenditure Analysis.* Washington, D.C.: Brookings Institution, 1968.

Schelling, T. C. (ed.). *Incentives for Environmental Protection.* Cambridge: The MIT Press, 1983.

Schultze, C. L. *The Public Use of Private Interest.* Washington, D.C.: Brookings Institution, 1977.

Settle, R. F., and B. A. Weisbrod. "Governmentally-Imposed Standards: Some Normative Aspects." Discussion Paper No. 439-77, Institute for Research on Poverty, University of Wisconsin, September 1977.

Slesin, L., and J. Ferreira, Jr. "Social Values and Public Safety: Implied Preferences Between Accident Frequency and Severity." Unpublished paper, Massachusetts Institute of Technology, September 1976.

Smith, R. S. "The Feasibility of an 'Injury Tax' Approach to Occupational Safety." *Law and Contemporary Problems* 38 (Summer–Autumn 1974): 730–744.

Smith, R. S. *The Occupational Safety and Health Act.* Washington, D.C.: American Enterprise Institute, 1976.

Spence, M., and M. Weitzman. "Regulatory Strategies for Pollution Control." In A. F. Friedlaender (ed.), *Approaches to Controlling Air Pollution.* Cambridge: The MIT Press, 1978, pp. 199–219.

Starrett, D., and R. Zeckhauser. "Treating External Diseconomies—Markets or Taxes?" In J. W. Pratt (ed.), *Statistical and Mathematical Aspects of Pollution Problems.* New York: Dekker, 1974, pp. 65–84.

Terkla, D. G. "The Revenue Capacity of Effluent Charges." Ph.D. dissertation, University of California, Berkeley, 1979.

Thaler, R., and S. Rosen. "The Value of Saving a Life." In N. Terleckyj (ed.), *Household Production and Consumption.* New York: National Bureau of Economic Research, 1976, pp. 265–298.

Thorpe, J. J. "Epidemiologic Survey of Leukemia in Persons Potentially Exposed to Benzene." *Journal of Occupational Medicine* 16 (1974): 375–382.

Tietenberg, T. H. "Derived Decision Rules for Pollution Control in a General Equilibrium Space Economy." *Journal of Environmental Economics and Management* 10 (1974a): 3–16.

Tietenberg, T. H. "The Design of Property Rights for Air Pollution Control." *Public Policy* (Summer 1974b): 275–292.

Tversky, A., and D. Kahneman. "Judgment Under Uncertainty: Heuristics and Biases." *Science* 185 (1974): 1124–1131.

U.S. Environmental Protection Agency. *Assessment of Health Effects of Benzene Germane to Low-Level Exposure.* Office of Research and Development, No. EPA-600/1-78-061, 1978a.

U.S. Environmental Protection Agency. *Standard Support Environmental Impact Statement for Control of Benzene from the Gasoline Marketing Industry.* Draft report, Office of Air Quality Planning and Standards, Research Triangle Park, N.C., May 26, 1978b.

U.S. Environmental Protection Agency. *Benzene Emissions from Maleic Anhydride Industry —Background Information for Proposed Standards.* Emission Standards and Engineering Division, Research Triangle Park, N.C., No. EPA-450/3-80-001a, February 1980.

Viscusi, W. K. "Labor Market Valuations of Life and Limb: Empirical Evidence and Policy Implications." *Public Policy* 26 (Fall 1978): 359–386.

Weitzman, M. L. "Prices vs. Quantities." *Review of Economic Studies* 41 (October 1974): 477–491.

Whiting, B. J. "Regulatory Reform and OSHA: Fads and Realities." *Labor Law Journal* (August 7, 1979): 514–525.

Wilson, R. "Examples in Risk-Benefit Analysis." Paper presented at Conference on Advanced Energy Systems, Denver, June 1974.

Zeckhauser, R. "Procedures for Valuing Lives." *Public Policy* 23 (Fall 1975): 419–464.

Zeckhauser, R. "Concepts for Measuring Risks and Benefits." Appendix A to *Principles and Processes for Making Food Safety Decisions.* Report of the Social and Economic Committee of the Food Safety Council, Washington, D.C., 1979.

Zeckhauser, R., and A. Nichols. "The Occupational Safety and Health Administration: An Overview." In *Study on Federal Regulation*, Committee on Governmental Affairs, U.S. Senate, 1978, appendix to vol. 4, pp. 161–248.

Zeckhauser, R., and D. Shepard. "Where Now for Saving Lives?" *Law and Contemporary Problems* 40 (Autumn 1976): 5–45.

Index